The Grand Delusion

THE GRAND DELUSION

What We Know But Don't Believe

Steve Hagen

Wisdom Publications
199 Elm Street
Somerville, MA 02144 USA
wisdomexperience.org

Library of Congress Cataloging-in-Publication Data
Names: Hagen, Steve, 1945– author.
Title: The grand delusion: what we know but don't believe / Steve Hagen.
Description: Somerville, MA, USA: Wisdom Publications, 2020. | Includes index.
Identifiers: LCCN 2019055445 (print) | LCCN 2019055446 (ebook) |
    ISBN 9781614296782 (paperback) | ISBN 9781614296799 (ebook)
Subjects: LCSH: Knowledge, Theory of. | Belief and doubt.
Classification: LCC BD161 .H135 2020 (print) | LCC BD161 (ebook) |
    DDC 121—dc23
LC record available at https://lccn.loc.gov/2019055445
LC ebook record available at https://lccn.loc.gov/2019055446

ISBN 978-1-61429-678-2   ebook ISBN 978-1-61429-679-9

26 25 24 23 22
6 5 4 3 2

Cover design by David Henry Lantz. Interior design by Tony Lulek.

Please visit fscus.org.

In remembrance of Jean

First,

> *Mountains are mountains and rivers are rivers.*
>> *This is delusion.*

Next,

> *Mountains are not mountains*
> *and rivers are not rivers.*
>> *This is necessary, but still delusion.*

Then,

> *Mountains are mountains and rivers are rivers.*

CONTENTS

Part II: Grand Symmetry and Grand Delusion

ACKNOWLEDGMENTS

My sincere thanks to Cal Appleby, Robert Evans, Ryan Johnson, Steve Matuszak, Vincent E. Parr, PhD, Linda Stevenson, and Mary Sullivan, whose careful readings of the manuscript yielded improvements and corrections.

Thanks also to Elizabeth Anderson, Richard DeWald, Jean Forester, Maurice E. Hagen, Wayne Lewis, Anne Morrow, Norm Randolph, Peter Wilson, and Susan Zeman. Each contributed in their own way—through research they did on my behalf, in stimulating conversations, or, like ANYONE, by peppering me with intriguing questions.

My profound thanks to Jose Palmieri, who has assisted me in countless ways over these past twenty-plus years, and who created all the graphics that appear in this book (as well as the website that accompanies it). Jose also asked intriguing questions and shared many stimulating conversations with me.

My special thanks to Scott Edelstein, my literary agent, editor, and friend of many years, with whom I have also had countless stimulating conversations, many of which also made their way into this book. Without Scott's efforts and know-how, none of my writings would likely have made their way out into the world.

INTRODUCTION:
BIG UNSETTLED QUESTIONS

Why is there Something rather than Nothing?

Humans have grappled with some version of this existential question for millennia—yet today we seem to be no closer to answering them than were our forebears of hundreds of generations ago.

Does God exist?

What does it mean to exist?

What is mind?

What constitutes measurement? This is a big problem for physicists.

We could just move on from these questions—but what exactly is motion?

All of these fundamental questions—and many, many more—stem from a single error, a single unwarranted belief, a single Grand Delusion. Clarifying this Grand Delusion is the aim of this book.

Most of what follows is a dialogue between me and ANY-ONE: a naïve but earnestly questioning character who could be, well, anyone—a person of any gender, any age, who may be anywhere on spectrums of learned and benighted, pious and profane, serious and silly.

When what I point out causes you to want to cut in with an objection, it is my hope that ANYONE will speak for you.

Be patient, though. I won't always get to your question or objection when it first arises. But I'll likely circle back to it later on.[1]

Time Out!

Every now and then, the main text of this book will be interrupted—and augmented—by *Time Outs* such as this one, set in this fashion. These contain additional background or contextualizing information that directly relates to what is being discussed. I recommend not skipping over them.

Endnotes and several appendices provide additional important information that further expands upon what appears in the main text. These are not mere references and citations. Much of the depth of this book will be found here.[2] There is also a glossary of technical terms—not because the particular words listed there will be new to you, but because some terms will be used in non-standard ways.

One other thing: at times, I will repeat a key theme—yet every time a theme is repeated, we will go a little deeper and look at ever more subtle aspects of a topic. I invite you to greet each iteration as though meeting it for the first time.

A Note on Truth v. truth

Although ANYONE repeatedly fails to acknowledge it, there are two truths.[3] References to conventional or everyday truth—relative truths related to concepts, ideas, and objects—appear in lowercase. References to Absolute Truth are capitalized. This allows a critical distinction—which will become evident over the course of the text—to be made without further qualification. Much of our confusion stems from the fact that our everyday conceptual terms cannot refer to Ultimate Truth or Reality, but only to conventional, relative truths. More on this later.

References to Awareness of Truth and Reality, such as *knowing* or *seeing*, are indicated in my remarks (but not in ANYONE's) through the use of *italics*. Conventional uses of these terms—without italics—refer to the knowing or seeing of thoughts, feelings, ideas, or objects in the conventional sense in which those words are generally used. This, too, will become both clear and familiar as you move through the book.

Bold text will be used for simple emphasis. ***Bold italics*** will be used for technical terms.

PART I

MIND, MATTER, MOTION, AND MUSIC

Questioner: Should we not seek for anything at all?

Huang-Po: By conceding this, you'd save yourself a lot of mental effort.

Questioner: But there can't just be nothing.

Huang-Po: Who spoke of nothing? You wanted to seek for something.

Huang-Po: This is It; as soon as you stir your mind, you miss It.

I. THE ULTIMATE QUESTION

What is the ultimate question?

For Bertrand Russell, the ultimate question was "Why is there something rather than nothing?"

When I first came upon that question as a young man, I immediately felt that nothing could be more profound than knowing the answer to it.

I quickly surmised, however, that there was no easy way to find out. How could we possibly know? Asking this question is like throwing it into a boundless void. How could we expect such a question to bounce back with an answer? It was as if, once launched, the question only ranged ever outward, propelled by its own inertia, never to return.

It was at once both unanswerable and utterly compelling. And for quite some time it haunted my thoughts like a spectre.

I listened to a lot of contemplative music in those days of my youth, and spent long hours brooding alone in the dark. I sometimes would listen to *The Unanswered*

Question by Charles Ives[4] when I was in this mood. This piece of music provided the ideal ambiance for serious contemplation of such a deep, all-pervading mystery.

Ives seemed to capture our ultimate predicament in his offbeat score. The piece opens with distant strings playing offstage in hushed, almost imperceptible harmonies. These slowly shift through expansive chords, evoking the timeless silence and immensity of space, the indifferent universe rolling on without end.

Then a lone trumpet calls out from the rear of the darkened hall (or, as I imagined, from a darkened wood or some other lonely outdoor place under a starry sky), as though asking a question. All the while, the indifferent strings roll on.

After the trumpet dies away, a quartet of flutes suddenly perks up, as if awakened from some long, primordial slumber. They seem as our forebears living long ago in the aboriginal dawn. Encountering the question for the first time, they give voice to our earliest response to the trumpet's call.

And still the strings maintain a distant backdrop of faint harmonies and shifting, airy, expansive chords.

The trumpet repeats. Again, the flutes revive, but now they focus on the question itself as the strings roll on.

Again, the questions: "Who are we? Why are we here? How did we get here? Where did we come from? Where are we going? Why is any of this happening at all? What is it all about?"

And repeatedly, the flutes, like innocent, eager children, attempt an answer.

Again, the trumpet sounds. And again, the flutes scramble to answer—again, without success.

Unrelenting, with each succeeding trumpet call, humanity's responses become ever more fragmented, agitated, and distressed—and further and further removed from any satisfactory answer.

Throughout it all, the muted, dispassionate strings roll on, utterly indifferent.

And this appears to be where human beings are fated to remain.

Except that we're not.

Seeing how and why we are not is the focus of this book.

2. SUBLIME OR TRIVIAL?

Eventually I realized that it wasn't just Russell (and Ives). Many great writers, thinkers, artists, scientists, theologians, philosophers, and composers have contemplated this question. But it was Gottfried Wilhelm von Leibniz who first wrote in 1714 (in German): "Why is there Something rather than Nothing?"[5]

Since then, innumerable thinkers—from Hegel to Russell to Nozick—have, like the flutes in Ives's piece, vainly sought a satisfactory response. Some have also commented on the question itself. Friedrich Wilhelm Joseph von Schelling, for example, called it "the final desperate question." And Martin Heidegger, who took it up as his central theme in *Being and Time (Sein und Zeit)*, described it as "dreadful."

Today, philosopher Derek Parfit of All Souls College at Oxford feels that "No question is more sublime than why there is a Universe: why there is anything rather than nothing." Yet others find the question wholly dismissible. Philosopher Bede Rundle, also of Oxford, in his

2004 book *Why There Is Something Rather Than Nothing*, argues that the existence of the universe "is not . . . a fact in need of explanation." Adolf Grünbaum expresses a similar sentiment in a 2008 *Free Inquiry* article, when he concludes that the question "should not engage your curiosity." To these thinkers, the existence of the Universe is both obvious and trivial, rather than a cause for urgent and profound inquiry.

As we shall *see*, however—and counterintuitive as it may first sound—we don't actually have Something rather than Nothing.

ANYONE: *Yeah, right. That's bananas!*
And who are you?

ANYONE: *Who am I? Who are you to make such a ridiculous proclamation?*
No one . . . in particular.

ANYONE: *So what makes you feel you can spout off like a fool? What you're saying is ludicrous!*
I understand. It seems bizarre. But in the pages that follow, we will *see* how and why this assertion in fact points to actual experience.

3. BREAKING OUT OF HABITUAL THOUGHT

ANYONE: *Your claim is complete nonsense, right on the face of it! Obviously, we don't have* **nothing!**

Obviously. But I didn't say that we have Nothing. I said that we don't have Something rather than Nothing. I'm saying that this is a false choice.

ANYONE: *That's completely off the wall!*

Yes. What I'm saying runs counter to our most basic assumptions about the World, about the nature of Reality, and about us. But seemingly outrageous statements can sometimes point the way out of what appear to be utterly perplexing quagmires.

History is full of such examples. Einstein had some of his biggest breakthroughs when he entertained the possibility that time is **not** constant, as Newton (and most others) had thought, or that light has a particle-like quality about it—even though, after 1803, everyone (including Einstein) "knew" it to be wave-like.[6]

Such breakthroughs have punctuated and changed history. When Aristarchus (in the fourth century BCE) postulated that the Earth moved around the sun, Aristotle and others dismissed the idea as absurd. And Aristarchus's idea remained absurd to almost everyone—until Copernicus heard about it 1,700 years later.

ANYONE: *Yes, but Einstein and Copernicus turned out to be right in the end. What you're advocating can't possibly be right.*
I'm not advocating a position. I'm only pointing to an erroneous belief on the part of nearly all of us.

History is replete with examples of how correcting one basic erroneous idea greatly expanded our understanding of the World. Almost invariably, such shifts in our thinking first appeared ridiculous or impossible to our contemporaries.

ANYONE: *What are you getting at?*
I'm pointing out that one small shift in our understanding can clear up all our confusion.

ANYONE: *You mean about the Big Question?*
Yes.

ANYONE: *Good luck with that. If it were **that** easy, it would have been done long ago.*

Who said it was easy? And who said it wasn't done long ago?

4. NEITHER EDGED NOR EDGELESS

I can provide an analogy.

ANYONE: *I'm listening.*

Imagine you're walking along a beach with someone from the Dark Ages. She tells you that she's been troubled by a gnawing problem. Pointing toward the horizon, she asks, "Does the earth just go on and on forever in all directions, or does it have an edge, out there somewhere?"

You tell her it neither goes on forever nor has an edge.

She frowns. "That's ridiculous. It's got to be one or the other," she says.

"The Earth is a sphere," you say, "or nearly so. So its surface is finite, yet it doesn't have an edge."

She laughs. "But look!" she says, pointing to the horizon. "It's flat. Look at that line out there—how straight and level it is. Look at how the water stretches away from the shore, on and on forever." She reaches down, throws up some sand, and says, "And I suppose on the other side of this sphere, sand falls up? And rain! Does that fall up, too? And am I supposed to believe that on the other side people walk around with their feet higher than their heads?"[7]

As she says this, you catch sight of an approaching ship on the horizon. You hand her your binoculars, quickly demonstrate their function, and have her take a look.

What she sees is utterly baffling to her. As she stands there, she clearly sees the ship slowly rising out of the water.

Because she has not yet digested the idea of a spherical Earth, the moment is surreal to her. "How can this be?" she asks you, her voice now trembling.

"The water's surface is curved," you tell her.

You explain that the Earth is a planet that rotates and orbits the sun. With the aid of a few pictures drawn in the sand, you show her how understanding these movements enables humanity to account for night and day, and the changing seasons.

It takes her some time and effort, but once she makes this simple shift in her understanding, she understands why the ship seems to rise out of the water. She is also eventually able to understand celestial phenomena such as the apparent motions of planets and stars, and lunar and solar eclipses.

This shift opens a door that, over time, allows her to begin to understand other phenomena that otherwise would appear far removed from her original question—things that would otherwise vex and perplex her. She is now poised to realize the great age of the Earth, for example, and all that this entails, with its implications for geology, biology, and evolution—as well as philosophy and religion.

More notably, the question that originally seemed so urgent and compelling to her—does the Earth go on forever, or does it have an edge?—*no longer troubles her at all*. She now realizes it's not a valid question. It only arose from her ignorance.

ANYONE: *So?*

I'm saying that the question "Why is there Something rather than Nothing?" is analogous to "Does the earth go on and on forever, or does it have an edge?"

I'm also saying that when we genuinely *see* this, so many other questions that vex us also clear up.

5. HOW WE PERPETUATE IGNORANCE

ANYONE: *Okay, so what are we supposed to figure out in order for everything to clear up?*

Don't worry about figuring out anything. It's not necessary to figure anything out, or to theorize, or even to look "out there." You need only to look at your assumptions, your unexamined beliefs.

ANYONE: *Huh?*

Think of the person from the Dark Ages. As long as she believed that the Earth is flat, she was preoccupied with imaginary concerns and irrelevant questions. In the same way, nearly all of us are captive to many erroneous beliefs about the World and existence.

ANYONE: *What you say about an uneducated person from the Dark Ages makes sense. But how is that analogous to us today? What do we not understand?*

Plenty. Stop me when you've had enough:

What is happiness and how do I find it?

Why do I quickly tire of things after I work so hard to acquire them?

What do I really want?

Why is there suffering?

What, if anything, is truly good?

How can we determine what is just or fair?

What is consciousness, exactly?

What is real?

What is true?

Does God exist?

What happens when I die?

Is there some purpose to life?

Do we have free will?

Is human existence meaningful or absurd?

How can I be sure of anything?

ANYONE: *All right, enough! And you're going to answer all these questions, I suppose?*
It's not necessary.

ANYONE: *Why not?*
They're FEQs. All of them.

ANYONE: *FAQs?*
Not FAQs—*FEQs. Flat-earth* questions.[8]

ANYONE: *What do you mean?*
They're all based on a false assumption, so they have very little to do with Reality.

ANYONE: *What false assumption?*
Substantiality. Our belief in solid, persisting things.

ANYONE: *As opposed to . . . ?*
Reality.

Consider quantum mechanics, the branch of science that deals with the behavior of atoms and their constituents—protons, neutrons, electrons, and so on.

According to the standard view of quantum mechanics, depending on how things are observed, these ostensible bits of matter—atoms and/or their constituents—can behave either as waves or as particles.

ANYONE: *Right, I've heard about that. But I don't see what the big mystery is.*
Have you really thought this through? When you turn on a switch and electrons flow through a metal wire—as when, say, you're in your kitchen operating an electrical appliance—these ostensible "things" appear to behave like particles. But if you pass "them" through a finely

spaced lattice, as you would when operating an electron microscope, it seems "they" behave like waves.

ANYONE: *Okay, so?*
So what are they? Waves or particles?

ANYONE: *They're both, obviously.*
Consider what you're saying. Sand grains on the beach are not at all like the waves lapping the shore—waves are spread out; particles appear point-like. Waves and particles are not at all alike. And why is there such a profound difference simply because of the type of observation one makes, or the type of job to which one applies the electrons?

Most scientists haven't a clue—because, like modern day flat-earthers held captive by subliminal ideas, they misinterpret what observable phenomena directly reveal.

ANYONE: *What do they reveal?*
Nonsubstantiality.

ANYONE: *Huh?*
And so, like any of us might, they continue as though electrons and other sub-atomic particles **are** Something.

ANYONE: *Well, duh! Of course!*

But this wave/particle enigma is only one of a virtually endless array of "mysteries" that would quickly disappear once we drop our deeply held substantialist view.

ANYONE: *What do you mean, "our substantialist view"?*
The belief that there actually **is** Something as opposed to Nothing.

ANYONE: *Are you saying that if we simply stop holding that view, all these mysteries would drop away?*
Right. In fact, they would never occur to us in the first place, any more than the flat-earther's questions trouble us today.

> **Confused by thoughts, we experience duality in life.**
> **Unencumbered by ideas, the enlightened see Reality.**
> **—*Huang-Po***

ANYONE: *But we can't just stop believing what's all around us. Besides, how are we supposed to do that?*
We can't do it by a simple act of will, to be sure. Substantiality does appear deeply compelling—and deeply intuitive.

ANYONE: *So how are we supposed to do it, then?*
By *just seeing*. That is, *just seeing* while refraining from mindlessly chattering to ourselves.

ANYONE: *You know, it's getting a little hard to take you seriously.*

Stick with me. Bit by bit as we explore, all of this will start to become clear. Or at least clearer.

ANYONE: *So you're saying that atoms and their constituents aren't actually real?*

No. I'm asking you to examine the idea that an objective thing can be both wave and particle. If you look closely enough, you'll see the mistake.

ANYONE: What mistake?

The tacit assumption that atoms and their constituents actually **are** things.

Time Out!

Here are a few items that beg for explanation if particles[9] are taken to be actual things:

(1) If particles were actual things, then it would be reasonable to assume that they have specific positions in space that change in time as they move. But quantum theory doesn't allow this. And, though there are instruments that seem to reveal tracks that these "particles" ostensibly follow, upon close scrutiny the "tracks" show themselves to be discontinuous, appearing not as unbroken straight or curved lines, but only as a series of, say, bubbles in a bubble chamber. At no

point is it obvious that anything moved through the chamber leaving such "tracks"—especially given the bizarre ways in which many of these phantom "things" would have had to move.

(2) If particles comprise matter, then it would be reasonable to assume that in a total vacuum there would be no subatomic activity. But quantum theory correctly predicts that a Geiger counter placed in a total vacuum will still click—i.e., detect "matter." In other words, "matter" cannot consist of whatever it is we think we mean by the term "particle."[10]

(3) If particles were real, objective things, then it would be reasonable to assume that they exist. But here's what happens when two observers view the same vacuum: If one person is at rest in relation to the vacuum, while the other is accelerating in relation to the vacuum, the former will see a cold vacuum, while the latter will see a warm gas of particles.

So, are particles real or are they an illusion? Are they Something or are they Nothing?

It would seem that such "difference" resides only in the minds of observers. In other words, the distinction arises as a purely mental phenomenon, not a physical one.

(4) If a particle is a physical thing, it should have the basic properties that belong to such things, such as energy, momentum, position, and so forth. But

particles can become entangled without showing any obvious material links among them, leaving "each particle" devoid of such definite properties. Only the entanglement itself, as a whole, exhibits such properties.[11]

If you believe in substantiality, it would serve you well to investigate whether there actually **is** Something substantial to build upon.[12] You might find that belief in substantiality inevitably leads to a host of apparent enigmas and contradictions. These appear in virtually every facet of our lives, not just in quantum mechanics.

As we'll see, however, all sorts of perplexing phenomena clear up when we abandon our deeply entrenched substantialist view.

ANYONE: *But isn't that science's job? To discover, study, and explain perplexing things?*
The problem is, regarding these big questions, science always comes up short.

ANYONE: *Why do you say that?*
Because all scientific inquiry is based on belief in substantiality.

ANYONE: *So?*
Well, as you implied, scientific inquiry is carried out to push back the frontiers of our ignorance. But belief in

substantiality actually kills all possibility that science will ever bring us to True Knowledge.

ANYONE: *What do you mean?*
In at least one important sense, science **perpetuates** ignorance; it doesn't whittle it away.

ANYONE: *Huh?*
Up front we're pressing into new territory, learning many new things as we go, much of which helps us to slough off mistaken notions from the past. That's fine. But what we don't understand is that, even as we let go of old, disproven beliefs, we keep manufacturing and holding newer, more sophisticated beliefs as we go.

ANYONE: *What's wrong with that?*
This isn't actual progress.[13] If we're interested in True Knowledge, we need to set aside these new, more sophisticated beliefs, too.

This process of endlessly hypothesizing, rather than *just seeing*, keeps us in the dark as to what's really going on.

For example, scientists have no idea why or how making an observation changes a potential (an apparent Nothing) into a manifest physical object (an apparent Something).[14] Neuroscientists are utterly baffled as to how the brain prepares to make a voluntary movement, even before the person is aware of deciding to move.[15]

Cosmologists scratch their heads because it seems that the constants of nature might not be constant after all.[16] Science gives us unlimited examples of how our expansion of ignorance outpaces our expansion of conventional knowledge. Science inevitably creates many more mysteries than it solves.

ANYONE: *Are you saying that science is useless?*
Far from it! Science is very useful in dealing with all kinds of practical concerns and technical issues—improving health, building bridges, tracking sea level rise, and so on. But science, impeccable though its method is, is simply incapable of bringing us to a complete understanding of Reality.

ANYONE: *But you can?*
I can't make you *see* what you already *know*. But in this book, I point out a few things that might help you *notice* what you've been overlooking.[17]

6. SUBSTANTIAL CONFUSION

ANYONE: *It's hard to believe that life's most profound enigmas can be done away with through a simple alteration of one's view.*

Bear with me for a while. You'll soon see that we can't make a sound argument for substantiality, and that any attempt to do so only invites paradox and confusion.

ANYONE: *So, someone who believes in substantiality is . . . what? A substantialist?*

Exactly. A substantialist is anyone who subscribes to the doctrine that what exists is matter, or matter and energy. They could be called materialists as well, but that term also refers to people who love to acquire money and goods and power, so the term *substantialist* is less potentially confusing.[18]

Substantialists also believe that even mind and consciousness are entirely the result of matter.[19]

ANYONE: *Well, yeah! Who doesn't believe that? I suppose there might still be some folks who argue for both mind and matter equally, but I don't think serious scientists or philosophers hang onto that view anymore.*

You mean Cartesian dualism—that the physical body is something quite distinct from the mind, and vice versa.

ANYONE: *Right. I mean, you can't deny that the world is basically, if not entirely, material. And, as neuroscientists have pointed out, the mind is what the brain does.*[20] *I mean, the brain gives rise to mind and consciousness.*

Don't be so sure. Scientists who believe this are at a total loss to support the claim. It's called the "hard problem," and with good reason, as we'll see.

This kind of thinking imagines all kinds of stories. It imagines that, as elements evolved, matter came together in just the right way to create life. It imagines that we are essentially protoplasmic machines. It imagines that we may someday be able to merge with robots or computers and become fully bionic, replacing our parts with hardware as needed. It imagines that machines themselves will eventually be made conscious.

ANYONE: *Well, according to what I've read, you'll soon be able to see projections in your eye, just like on a computer screen. Doesn't this indicate that everything, including*

mental functions like consciousness, is materially based?
Just like what goes on in a computer.

If you read more widely across the spectrum of scientific inquiry, you'll find plenty of evidence that mind, consciousness, and even life itself are unconstructed and without origins. You simply can't demonstrate that brains create mind and consciousness. In fact, along with life, you can't even show that mind and consciousness are created at all.

ANYONE: *What are you talking about?*

Well, for example, given the abundance and ubiquity of the elements necessary for life throughout the universe,[21] there appears a continuous spectrum, running from what we call non-life to what we call life, with no threshold or barrier between the two. (Viruses, for example, have long been viewed as life in some ways but not in others; they can appear to remain indefinitely in crystalline form and yet "spring to life" when conditions become favorable.) But this is still to see things in gross terms.

What I'll be pointing out in this book is far subtler. As we'll see, life, mind, and consciousness can be *seen* as coming along with the Universe Itself, as a Whole, rather than as phenomena emerging from the evolving complexities of matter.

The fact is, if we try to maintain our substantialist view, we can have no conceptual clue as to what life, mind, and consciousness actually are. Or even if it makes sense to say that they have origins.[22]

And here's a related conundrum. We used to believe the World was made of matter and energy (two forms of essentially the same thing—whatever that might be—given the famous formula $E = mc^2$). But, according to current cosmological theory, only 4 percent of the universe is made of matter and energy. The bulk of it, 70 percent, is so-called "dark energy," and the remainder, 26 percent, "dark matter." (Though "transparent matter" might be a better term for it. Given that it doesn't interact with light, we look right through it without seeing anything.) At present, virtually no one, including the most devout materialist, has any clear idea what dark energy and dark matter are.[23]

Of course, materialists don't know what "regular" matter and energy are, either. Though physicists define energy as "the ability to do work," this only tells us what it does. It says nothing about what it is, or how matter is somehow a form of it—whatever it is.[24]

ANYONE: *Wait. You're saying that we're still basically clueless about everything that makes up the universe?*
Yes. And scientists readily admit to 96 percent of this ignorance.

The reason they suspect that dark matter is "out there" is because we can see large-scale gravitational effects on ordinary matter that are much greater than can be accounted for from the presence of matter that we can see—that is, the matter that makes up all the material things we're familiar with, including our own bodies. All the galaxies appear to be surrounded and infused with this dark matter. Thus, some scientists think, it must pass through our bodies continuously. Yet, on the small scale, it does not appear to interact with photons, or even with the subatomic particles that ostensibly make up our bodies, or with any other small-scale aspect of the so-called material world we seem to find around us.

> **Astronomers have long inferred that most of the material in the universe is invisible, existing as mysterious dark matter. But a new study suggests that most ordinary matter is hidden as well.**
> *—Andrew Grant*[25]

As for dark energy, like dark matter it's only describable by its effects—namely that space appears to be expanding. In fact, the universe appears to actually be falling outward.[26] Scientists don't know why, but they infer that dark energy must exist, since there is no other way to explain the observed fact that the universe appears to be expanding at an accelerating rate, rather than slowing its expansion under the effects of gravity.

Thus, science, given the effects of dark energy and dark matter, admits to having no basic understanding of what constitutes 96 percent of the physical universe.

ANYONE: *But there's still that 4 percent.*
Yes, but most of the material that supposedly comprises the remaining 4 percent seems to be missing as well. For all we know, it's all "ordinary matter"—yet most of it doesn't show up.

So, it would seem that we haven't a clue as to what makes up 96 percent of the physical world, and only a vague idea about the remaining 4 percent. Our ignorance about the substantiality of the material world, let alone how to account for it, is fairly complete.

Just to drive the point home, we can throw in for good measure this fact: physicists have found that 90 percent of our own ostensibly physical mass is from virtual particles that zip in and out of reality so fast that they don't stay around long enough to qualify as actual, substantial particles. And it appears that all the mass in the universe seems to have come about in this way, from virtual particles.[27]

Beyond all of this, scientists note that there are two forms of energy—one we call positive, and one we call negative.[28] It appears, however, that if we add up all the matter and energy in the universe—including dark matter and dark energy—the total would sum to zero. In other words, if we could get all this apparent Something-ness together into one place all at once, it would add up to Nothing.

ANYONE: *But I thought you said that we don't have nothing.* Obviously. And by the same token, neither have we Something. So, something other than the mere totaling up of matter and energy must be in play.

ANYONE: *Are you saying that everything in the universe has no substance?*
Do you hear the contradiction in your question?

ANYONE: *No.*
The very word "everything" imagines substance—and an array of separate things that can possess it.

ANYONE: *Hmm.*
Stop and notice something else here. We've looked at a slew of inadequacies that seem endemic to what we call matter and energy. We don't understand what they are, and we can't even seem to find them as objectively real. Yet Mind has not been affected in any way whatsoever by these inadequacies. While matter is vague, undefined, and mostly or entirely missing, Mind is immediate and *here.*

ANYONE: *But the brain is made of matter!*

Like a flat-earther, you're still hanging on to the unfounded belief that brains produce consciousness. Actually, you've got this picture completely backward.

ANYONE: **WHAT**?

Let's take a closer look and *see*.

7. A UNIVERSE OF MINDSTUFF

Consider this low wooden coffee table in front of us. Its top is exactly one meter square. I measured it.

ANYONE: *Okay. That looks about right.*
We can feel its hardness and see its color and shape. To us the tabletop appears to not move. It appears to just sit there, persisting as itself from moment to moment.

ANYONE: *Right.*
But as we zoom in for a closer look, that isn't what we find at all.

Imagine projecting this tabletop, full scale, onto a meter-square screen in front of us.

ANYONE: *Okay. I'm picturing it.*
Now let's blow up a square millimeter of the tabletop to fill the screen. At this magnification—one thousandth of a meter (one millimeter, or 10^{-3} meters) on a side,

expanded to one meter—we have enlarged the tabletop itself a million times, and the wood fibers loom large.[29]

Now let's move in the same amount of magnification once again. Now one-millionth of a meter (one micron, or 10^{-6} meters) is expanded to one meter. We can see the membrane walls that enclose cell nuclei within the wood's cellular structure.[30]

Let's keep moving in further. At 10^{-8} meters linear we can observe twisting strands of DNA. At one 10 billionth of a meter (10^{-10} meters, or one angstrom), we have reached the scale of the atom.[31]

Now, remember, all we've done in this thought experiment is magnify the object. We've done nothing to alter appearances other than that. But once we're in close enough to "view" the table's atoms, we start to notice that things appear a bit strange, at least if we also try to maintain our substantialist view. The atoms that we say "make up the tabletop" don't exhibit all the properties that we normally attribute to everyday entities such as tables, plants, planets, and people. They begin to appear indeterminate in a variety of ways.

ANYONE: *Such as?*
For starters, they appear as empty space.

ANYONE: *Yeah, mostly. At least that's what scientists say. Each atom has a shell around it where the electrons whiz around, and a solid nucleus at its core. That's where most of its mass is. I picture the electron whizzing around so fast that it makes a kind of surface . . . .*

No. That's a commonly held misconception based on early models of the atom, but any physicist will tell you this isn't even remotely what goes on.

First of all, the atom has no "surface" as such. It's not at all like, say, the surface of a billiard ball—at least, as we think of it. There's nothing material about it. It is, rather, a magnetic force, a push (or pull, as the case may be), as when you try to press two bar magnets together while their poles, N and S, are aligned like to like.[32] In this configuration the magnets repel each other. The closer you try to press them together, the stronger the repelling force. You can feel it, as if you were pressing two surfaces together, even though only empty space appears between the magnets. It's only these repelling pushes that give the impression of a surface.

This is what we're talking about here. Electromagnetic forces define the "surfaces" of everything we experience materially.[33] In fact, it's just such magnetic forces that give us our naïve impression of materiality.

So, when it comes down to the "surfaces" of atoms, it's purely pushes and pulls. No one has ever actually found any materiality there.[34]

ANYONE: *Where?*
Wherever an "electron" is supposed to be.

ANYONE: *You mean in the shell.*
Remember, the shell isn't a material object. It's a general location determined by force fields. As for the electron, there doesn't seem to be any "it"—any electron—there. Just the defining limit of what we're calling the atom. The "electron" itself would instead seem to be what physicists call a "point particle."

ANYONE: *A what?*
A point particle.

ANYONE: *You mean like a mathematical point?*
Yes. Without size or dimension.[35]

ANYONE: *Hmmm...*
So, the shell of an atom has no substance, the electron has no substance—and together they surround and enclose empty space.

ANYONE: *What about the nucleus? That's where nearly all of the mass of the atom is.*
Good question. Let's move in for a much closer look by expanding the atom's nucleus to the size of, say, an orange.

We've now left our tabletop image far behind, given the scale we're now talking about. How large a volume do you suppose the shell of this scaled-up atom would enclose? For simplicity's sake, let's say we're talking about a hydrogen atom, where there is only one subatomic particle in the nucleus—a proton—with one electron defining the atom's shell.

ANYONE: *I don't know.*

Picture taking our orange-sized nucleus into the center of a city of three or four million people. The shell of this atom—which, remember, is not a material object, but a locally defined force—would easily enclose the entire metropolitan area. And not just from side to side, mind you. We're talking about a sphere that would extend miles above the surface of the Earth, as well as miles below—a volume that size compared to an orange. That's a rough comparison of an atom to its nucleus. The rest is "empty space."

So, when scientists point out that matter is mostly empty space, that's what they're talking about. Even if the subatomic particles **were** solid entities, an atom is still almost entirely empty space.

But we're not done. It's not correct to talk about this nucleus as an actual object—like a billiard ball or an orange. Remember, we only used the image of an orange as a reference to relative size. But you are correct.

It is the nucleus that carries virtually all of the mass and weight—the "substance"—of this atom.

But is it a solid thing? Is it even there?

ANYONE: *Isn't it?*
What makes up the nucleus?

ANYONE: *It's made of quarks, isn't it?*
Isn't it nice how, when we can toss out a label, we feel like we've gotten hold of something? So, what's a quark?

ANYONE: *I don't know.*
How big are they?

ANYONE: *I have no idea.*
Do you suppose they occupy the whole of the nucleus, to give it all that weight and heft?

ANYONE: *Don't they?*
Again, let's say we're talking about the nucleus of a hydrogen atom—one proton. A proton is made of three quarks—two "ups" and a "down," as they are labeled.

ANYONE: *Okay.*
But can you ever have a single quark by itself?

ANYONE: *I don't know. Can't you?*

Not for long, it seems. But let's just poke inside this orange-sized proton nucleus to see if we can find these quarks.[36]

ANYONE: *What do we see?*

Nothing.

ANYONE: *Nothing?*

Just space.

ANYONE: *Maybe they're too small. Can we blow it up some more?*

Why not? This is just a thought experiment, so let's blow it way up this time.

ANYONE: *Yeah! Really big!*

Okay. So, how about we expand our orange-sized nucleus to the size of planet Neptune's orbit around the sun?

ANYONE: *Wow! How big is that?*

The diameter of the sphere we're talking about is more than five and a half billion miles across. This is large enough to hold not only the sun and all the planets, but the orbits of all the planets as well. And remember, we're not talking about a linear dimension, or even a circle. We're talking about a sphere.

So, now let's go in once again and look around this ultra-expanded sphere to see if we can find these elusive quarks.

ANYONE: *And what do we see?*
Nothing.

ANYONE: *Nothing?*
Nothing.

ANYONE: *How can that be?*
One cosmologist wrongly estimated that at this scale a quark would be expanded to about the size of a virus. But what does that mean compared to the volume of a sphere 5.5 billion miles in diameter? Bear in mind that this "sphere" we're talking about is actually the "size" of a proton.

ANYONE: *No wonder we can't see them!*
In fact, **they**—the quarks—are point particles, too. They have no dimension, and no size, just like the electron. Which is why we couldn't really find the electron, either, remember? They're like mathematical points. So, forget the image of a virus. There's nothing of any "size" there at all.[37]

Given all this, are we still talking about material existence here? Or is it time to admit that we've wandered into the realm of mindstuff? And, if so, what line did we cross, and when did we cross it?

ANYONE: *Okay. Maybe so, but this all seems rather abstract, doesn't it? I mean, we don't really experience anything like this in real life.*
On the contrary. *Every physical experience you've ever had is comprised entirely of such stuff.*

Remember, we did nothing arcane or hard to understand here. We simply magnified our everyday view of the physical world. We didn't alter or remove anything. All we did was move in for a close look—and all materiality revealed itself as mindstuff.

> **Physics is the study of the structure of consciousness.**
> **The stuff of the world is mindstuff.**
> **—Sir Arthur Eddington**

ANYONE: *Okay. But in everyday life, we never actually see anything like what we imagined in our thought experiment.*
Actually, we don't need to go to the subatomic level to reveal material objects as sheer mindstuff.

ANYONE: *Give me an example.*
Consider the paradox of the pile.

ANYONE: *The what?*

If you keep removing beans from a pile, at what point is the pile no longer a pile?

ANYONE: *Yeah, but that's . . . .*

This is only a paradox, of course, if we think "a pile" pertains to some intrinsically real thing, rather than to a concept—a mental object.

ANYONE: *Exactly.*

Yet the "pile paradox" applies to everything we believe exists in the physical world. If a hair drops from your head, are you still you? In the span of twenty-four hours, nearly all the cells of your pancreas are replaced. Materials are continuously flowing in and out of your body. How much replacement is required for you not to be you? Is this even a sensible question?

ANYONE: *What are you saying?*

"You" are no less a vaguely defined mind object than a pile of beans or a quark.

At the scale of the quark, this ever-present vagueness becomes abundantly apparent. But if we look closely and honestly at our everyday objects, and don't cling to any belief or assumption, we see exactly the same nonsubstantiality.

8. NO NEED TO EXPLAIN EVERYTHING

Science provides beautiful and precise descriptions of physical phenomena. And that's useful and important. But in the end, science doesn't really explain anything.

ANYONE: *Of course it explains things!*
Does it? Consider something as ubiquitous as magnetism. You can read numerous books on the subject and study it for years, yet nowhere is magnetism ever explained. Only described. Ultimately, we have no idea what it is or why it is. It's a force that we can measure and describe, but not explain.

ANYONE: *All right, point taken. So, what's* **your** *hypothesis about magnetism?*
Hypothesizing won't do at all. Not if we want to address the Big Question of why, say, there appears to be Something rather than Nothing.

ANYONE: *What do you mean?*

Theories and hypotheses generate more questions than they answer.

So, instead of formulating endless hypotheses in vain attempts to explain what's going on, we need to take a direct look at what we're ignoring.

ANYONE: *But creating theories and hypotheses is a way of probing deeper into the mystery.*

What mystery?

ANYONE: *All of what you've been going on about!*

There's no mystery here. As functional and useful as it is for us to come up with hypotheses "explaining" things, that is simply no way to go about resolving the Big Question.

When it comes to waking up to Reality, such methods take us in the opposite direction from where we want to go. Hypothesizing only expands "the mystery" by increasing the scope and depth of what we ignore.

Simply put, if our concern is to bring our ignorance to an end, then we ought not to ignore it.

ANYONE: *Ignore what?*

Our ignorance. We ought not distract ourselves from it.

ANYONE: *What do you mean?*
We need to *know*—to get clear about—what we don't know.

ANYONE: *Isn't that what I've been saying? The mystery is out there! And we're probing it!*
No. We've done just the opposite. We keep looking "out there" and supplying hypotheses.

ANYONE: *That's where the mystery is.*
What you're ignorant of is right *here*, not "out there."

Stop searching and hypothesizing. Instead *just see* what's already *here*.

9. NO STAND-IN FOR REALITY

So far, we've only glanced at the tip of this iceberg. Beyond what I've already pointed out, there are countless other aspects of this supposed world of Somethingness that ought to give us pause.

ANYONE: *Such as?*
There's no philosophical underpinning whatsoever for the mathematics of motion. Or time.

As we look into them, you'll see that both motion and time, as we imagine we experience them, are delusions stemming from our conviction that the world is fundamentally material.[38]

ANYONE: *Give me a break!*
Here's something else: most of us haven't a clue as to what constitutes mind or consciousness. In fact, many among us don't even have a clue that we don't have a clue.

ANYONE: *Could you explain, please?*
If we would only drop our substantialist view—in this case, our infatuation with brains—we would immediately cut through confusion on all fronts. When we *look* carefully, ultimately there doesn't appear to be any material basis for consciousness at all.

ANYONE: *Of course there's a material basis for consciousness! As I said, the mind* is *what the brain does!*
If you could suspend that particular foregone conclusion of yours . . .

ANYONE: *Mine and everybody else's.*
. . . and carefully attend to actual experience, you'd *see* directly for yourself that it doesn't hold up.

ANYONE: *How are you going to prove that?*
I'm not out to prove anything. I'm just pointing out an unwarranted assumption that's constantly made by nearly everyone.

ANYONE: *And the alternative view you're proposing is . . . ?*
I'm proposing that we drop our substantialist view. Which is not an invitation to take up yet another view.

I'm not positing a different idea or view. I'm only advocating that we drop a basic assumption almost all of us make and live by, and *see* what happens. This process itself will supersede all theories, all views, and all ideas.

ANYONE: *Huh? How?*

Consider how theories work. The power of a theory lies in the ratio of how much it clarifies versus how much it assumes. A theory that accounts for a great deal without assuming very much is far more powerful than a theory that accounts for little but assumes a lot.

Darwin's theory of what he called "natural selection" has been called the most powerful of all theories, because it provides an organizing principle for all of biology—as well as geology, astronomy, cosmology, computer networking, and many other sciences. All that is observed in these disciplines is made sensible simply by assuming natural selection. Thus, in Darwin's case, the ratio of what is made understandable—a lot—to what has been assumed—very little—is extremely high.[39]

On the other hand, as an example of an extremely weak theory, the competing theory of Intelligent Design explains virtually nothing (though it purports to explain as much as Darwin's), while assuming a great deal.

What we're doing here, you could say, lies beyond the Darwinian end of this spectrum. In fact, it lies beyond theorizing of any kind. Instead it examines what we unwittingly assume. This doesn't require that we theorize about anything—only that we *just see.*

ANYONE: *What's your point, then, if you're not explaining anything?*
I'm simply calling *attention* to the fact that substantialism is itself an idea—a concept—and nothing more. As such, it's not, and cannot be, a stand-in for Reality.

Yet we use it as though it were the be-all and end-all of Reality—not only in science and philosophy, but in all human affairs, including religion, politics, business, and every other aspect of daily life.

ANYONE: *How can everyone be wrong about reality?*
Good question. We'll *see* how in the pages that follow.

10. WHAT WE'RE MISSING

ANYONE: *It seems to me like you're saying that nothing exists.*[40]

No. What I've said is that, if we look carefully, we can *see* that we don't actually have Something rather than Nothing. We can't legitimately say "it is," no matter what the word "it" refers to.

ANYONE: *So, in effect, you're saying, "It is not."*

No. What you're appealing to is called the law of excluded middle, which demands that every meaningful entity, concept, or statement must be either true or false. Or, as you're applying it here, it demands that a thing must either exist or not exist.

ANYONE: *What other possibility is there?*

The law of excluded middle is one of three so-called "laws of thought" formulated after Aristotle more than two thousand years ago. The other two are the law of contradiction and the law of identity.

As I said, the law of excluded middle asserts that a thing, an idea, or a proposition must be either true or false. The law of contradiction states that a thing, an idea, or a proposition cannot be both true and false. The law of identity basically states that a thing is what it is. These "laws," however, are not universal, but merely principles of conceptual thinking. They describe ways in which we **must** conceive of our experience if we would make sense of it conceptually.[41]

But, as we'll see, they don't actually describe or reflect Reality.

ANYONE: *Or so you keep claiming. Does anyone besides you say this?*

By the nineteenth century, some mathematicians began to discredit the law of excluded middle. Its total undoing finally came in 1931, with Kurt Gödel, who proved the possibility of "entities" that neither are nor are not.[42]

Gödel came along just in time to shore up quantum mechanics—or so it may have seemed at the time—which hinted that there are all sorts of "things" in the physical world that do not belong to either the "it is" or the "it is not" category.[43] For example, electrons, positrons, quarks, and all other so-called "point particles," seem to be "there" at times—yet we can't pin them down. As we've seen, "they" are dimensionless.

And then there are virtual particles that zip in and out of reality before they can be deemed "real." Yet, as we saw, they make up the bulk of the mass of the physical universe, including your body.

So, regarding most[44] of what we call "the physical universe," we can't legitimately say "it is not" any more than we can say "it is."

What we are ultimately left with, in other words, is neither Nothing nor Something.

ANYONE: *That's absurd! I mean, that might be the case for subatomic particles or some arcane piece of mathematics that almost nobody understands, but such abstraction doesn't have any bearing on our everyday experience.*
Be careful about what you call abstract and what you regard as substantial. You've gotten these everyday notions completely turned around.

ANYONE: *Me and everyone else!*
Getting you to *see* this is going to take further examination, so let's press on.

Let's look at what we commonly mean when we use the word "nothing."

We typically use it to connote an object or concept— often a lack or an absence—as in "nothing in particular," "nothing of value," "nothing meaningful," or "nothing we can pin down, clearly define, or get hold of."

But we're still referring to something. So long as there's something being referred to—an image, sensation, thought, or dimension—the "nothing" that we're speaking of is still a something. At the very least, it's an idea, a mental object.

This, however, is not the Nothing-with-a-capital-N that Leibniz was wondering about when he asked, "Why is there Something rather than Nothing?"[45]

ANYONE: *What do you mean?*
He was referring to Nothingness.

ANYONE: *What's the difference?*
True Nothing would involve no sound, color, feeling, dimension, or thought. True Nothing would also involve no **lack** of sound, color, feeling, dimension, or thought. With True Nothingness there would be no consciousness, no Awareness. There wouldn't even be a blankness, for that is still something.

True Nothing cannot be thought or spoken of; otherwise it would be a something. Nor could there be someone there, such as Leibniz, or us, to contemplate this Nothing.

Clearly, then—as you yourself have observed—we don't have Nothing. There is never Nothingness. Immediate, direct experience bears this out.

ANYONE: *Agreed. So . . . ?*

Leibniz was asking, "Why is there anything at all?" He asked this because he devoutly believed, as most people do, that we have Something—that there actually **is** Something "out there." The possibility that substantiality itself is an illusion never occurred to him.

But, as we have seen, we don't have Nothing, and we don't have Something. We can't say "it is," and we can't say "it isn't." Reality is without substance, but It's not Nothing.

ANYONE: *So what* **is** *reality, then?*

Illusion.

ANYONE: *How's that, again?*

Reality is Illusion.

ANYONE: *But if reality were an illusion, it wouldn't be . . . .*

I didn't say Reality is an illusion.

ANYONE: *You just did!*

I said Reality **is** Illusion.

ANYONE: *But if reality were illusion, it wouldn't be reality!*

Only if It were an illusion of Something.[46]

II. *SEEING* IS NOT BELIEVING

ANYONE: *So, what exactly are you saying?*
Illusion is devoid of any intrinsic substance beyond mere appearance. Yet, though Reality is nonsubstantial, Reality nevertheless **appears** substantial. That's the Grand Illusion.

The fact is, **all** phenomena are devoid of intrinsic substance—and, hence, all are illusory.

ANYONE: *Hmm.*
What would you say is the deepest and most pervasive belief among human beings?

ANYONE: *I don't know. Belief in God?*
Belief in substantiality.

ANYONE: *But substantiality isn't a belief! It's reality!*
Thanks for making my point.

ANYONE: *Well, even if what you're saying is true, people need **something** to believe in.*

Actually, we don't. So I'm not encouraging belief in anything here. In fact, belief is the bane of humanity.

What I'm pointing out is simply a matter of *seeing*.[47]

ANYONE: *But isn't seeing believing?*
When referring to the seeing of concepts, images, and physical objects, yes. Believing is conceptual. More than that, it's imagining that our conceptualized versions of the World are True.

Seeing, on the other hand, is "just" *seeing*, without any conceptual overlay. It's pure perception. It's wordless. It doesn't involve concepts.

ANYONE: *Okay . . . but I still don't see what you're getting at.*
Belief—any belief—precludes Knowledge, or Awareness of the direct experience of Reality.

ANYONE: *Huh? How?*
When we are caught up in our indirect, conceptual **retellings** of direct experience—i.e., when we are caught up in belief—we ignore actual immediate, direct, perceptual experience.

Instead, our minds are saturated with ideas, theories, conjectures, concepts, opinions—all of which we accept as Real, both emotionally and intellectually.

For the most part, we live out of these mind objects. We become so thoroughly engrossed in our abstract mental constructs that we're completely unaware of how they're all made up. Thus, we become caught up in full-blown, non-stop delusion.

Unless we *wake up* to what we're doing—unless we *attend* to actual, immediate, direct experience—we remain unwittingly stuck in belief, in delusion, in a world of pure mental fabrication.

ANYONE: *Hang on. I've read some philosophy. What would you say, then, to Jacques Derrida, who said there is nothing without signs? Or what about Ludwig Wittgenstein, who claimed that all he knew is what he had words for. What would you have said to him?*

Actually, what Wittgenstein had words for is all that he conceived, rather than what he *knew*.

Words and signs are by nature concepts—mind objects, mental creations. What we *know*—such as the immediate sensing of taste or sound—is perceptual, never conceptual.[48]

ANYONE: *I don't understand the distinction you're making. How exactly do the conceptual and the perceptual differ?*

The conceptual has to do with what we **make** of experience; the perceptual has to do with immediate, direct experience itself.

Time Out!

Here is an illustration of how the conceptual differs from the perceptual. In his book *An Anthropologist on Mars*, Oliver Sacks relates the story of Virgil, a man who was blinded in infancy but had his sight surgically restored in middle age.[49] Sacks writes that "after being blind for forty-five years—having had little more than an infant's visual experience, and this long forgotten—there . . . was no world of experience and meaning awaiting him [once the bandages were removed from his eyes]. He saw, but what he saw had no coherence." Sacks continued, "Everyone, Virgil included, expected something simpler. A man opens his eyes, light enters and falls on the retina: he sees. It is as simple as that, we imagine."[50]

But what Virgil later reported, according to Sacks, was that, once the bandages were removed, "he had no **idea** [my emphasis] what he was seeing." In other words, he had no ready-made concept to interpret the sight. While there was apparent light, movement, and color, it was "all mixed up, all meaningless, a blur." It was only when Virgil's surgeon spoke to him that he realized that the "chaos of light and shadow was a face—and, indeed, the face of his surgeon."

Virgil's confusion was not something to overcome easily. In fact, Virgil never did get the hang of it. Under-

standing distance, seeing depth, and delineating out-
lines remained constant challenges for him.

But it's not just in such extreme cases where we
can parse conception from perception. In a footnote,
Sacks also mentions that "it has been reported that if
people who have lived their entire lives in dense rain
forest, with a far point no more than a few feet away,
are brought into a wide, empty landscape, they may
reach out and try to touch the mountaintops with their
hands; they have no **concept** [again, my emphasis] of
how far the mountains are."[51]

Simply put, conception, as useful as it might be, is no
substitute for perception. So, if we're interested in *waking*
to Truth, we need to stop relying on what we think—i.e.,
conceive—at the expense of what we *see*—i.e., perceive.

ANYONE: *What do you mean?*
Our problem is that we find it terribly difficult to let go
of our cherished views.

ANYONE: *What problem are you talking about, exactly?*
The Big Problem—the one that arises in our minds, as
we noted in chapter 1, like the unrelenting sound of Ives'
trumpet.

ANYONE: *Huh?*

True Knowledge doesn't come to us through concepts or beliefs—whether they're primitive or modern, simple or sophisticated. In fact, it's through concepts and beliefs that we get caught in interpreting everything in terms of what we already believe. We see what we expect to see, not what is actually going on. This is not Knowledge.

The wise reject what they think, not what they see.
The foolish reject what they see, not what they think.
—*Huang-Po*

ANYONE: *So, we're all a bunch of ignoramuses?*

I wouldn't put it that way. We're simply ignorers. We ignore actual, immediate, direct experience—what occurs **before** we form words or concepts or thoughts.

Indeed, we don't even notice that there is any "before." We're oblivious to actual perception because we're entirely occupied with our conceptualized interpretations, our stories about what is going on.

ANYONE: *You're not inside my head![52] How can you possibly know what I perceive? Granted, I might not pay attention to everything all the time. Like everyone else, I have to tune out a lot of stuff. Particularly routine things, like my drive to work, or the cup of coffee I hold in my hand as I walk to my desk each morning. We can't be tracking everything that's going on around us all the time. We'd be overwhelmed with sense data if we did. Anyway, one of the brain's primary functions is to eliminate or ignore sensory data that's unimportant. Right now, I perceive a tree outside my window, but I'm screening out the background noise and whatever else is in my peripheral vision.*

Actually, your examples make my point. You take "perceptual experience" to mean our experience of things like coffee cups or trees. But we never actually perceive coffee cups or trees, or any other formed object.

ANYONE: *Of course we do!*

If we slow down and carefully examine what's going on, we can *see* that this doesn't actually match immediate, perceptual experience. It only matches our **retelling** or **interpretation** of experience—what we've conceptualized.

ANYONE: *Now you've lost me.*

Perception refers to sensory experience. And we don't **directly** experience "coffee cups" through our senses.[53]

ANYONE: *Of course we do!*

No. We see shapes and colors; we feel hardness and warmth and weight and balance. We only **conceive** of a coffee cup after the sensory experience. A "cup of coffee" is a conceptualized experience, not a direct, perceptual one. It's an idea. *Seeing, smelling, feeling,* and *tasting* are perception.

We're conceptualizing when we form coffee cups, or people, or rivers, or ideas, or toothaches, or heartaches, or any other mind objects. And these conceptual constructs come "after" perception, so to speak.[54] We conceptualize entities from raw perception.

ANYONE: *So, what is actually perceived?*
Immediate experience.

ANYONE: *Yeah, but what **is** that?*
That's precisely our problem. We feel we have to give it form—conceptualize it, explain it, name it, describe it, categorize it. But it's best not to try to pin down perceptual experience.

In any case, the fact is that you can't.

ANYONE: *Why not?*
Because the moment you do, that's conceptualization—forming a mind object—not perception.

ANYONE: *I'm still not exactly clear about what you're saying.*
Perceptual experience itself can only be wordlessly realized—e.g., *knowing* the taste of orange juice—before any mind object gets formed. If you try to capture

that Knowledge in words, just note that whatever you describe falls short of actually *knowing* the immediate, direct experience. In other words, you will never be able to describe a perceptual experience in such a way that, say, one who has never tasted orange juice will *know* the taste from your description. They can only wordlessly *know* it by tasting it.

12. WHO DO WE THINK WE ARE?

ANYONE: *You're talking about experiencing just raw sense data, then?*

Not quite, but we can call it that for now.

ANYONE: *What do you mean?*

What you said is fine, as long as you realize that there's no actual data, as such.

ANYONE: *Huh?*

This is not as strange as it sounds. Slow down and reflect. Each "datum," if taken as an object in and of itself, would constitute a formed thing. Thus, we're back to conceptual experience once again, not pure, raw perception.

ANYONE: *So, when I perceive reality . . . .*

No.

ANYONE: *What do you mean "no"?*

Pure perception doesn't include "I."

"I" is a mental construct. "I" is never perceived—only conceived.

ANYONE: *What about Descartes? "I think, therefore I am."* Essentially, he's saying, "A concept, therefore substantiality."

The problem is, while there's the direct experience of thinking (i.e., perception), there's no direct experience of "I," the **object** of thought (i.e., conception). In other words, there's nothing about the assumed, conceived "I" that's directly perceived.

ANYONE: *Of course there is!*
What, particularly?

ANYONE: *I perceive myself.*
And what does "myself" refer to?

ANYONE: *Well . . . here I am! Here's my face; here are my arms and hands! Here's my body!*
You don't feel there's anything vague or abstract about what you're referring to?

ANYONE: *Here are my arms! How much more concrete can I get? What's vague or abstract about that?*
So you're saying that your face, arms, hands, and body are "you."

ANYONE: *Sure they are. Who else would they be?*
And "you" is "your body"?

ANYONE: *Right.*
Setting aside what we've already noted about the evanescent nature of matter, what if an arm were removed? Would it still be "you"?

ANYONE: *Technically, that would still be my arm, yes. But it would no longer be a part of me.*
Now you're speaking of ownership. There's a "you," and it owns an unattached, separate arm.

ANYONE: *Right.*
But would that arm **be** "you"?

ANYONE: *No, it would no longer be me, any more than my shirt would be me. It would only be me if I possessed it as a part of my body. If I lost my arm somehow, it would no longer be me, because I'm still here, and the arm is somewhere else. The arm was only mine while I had it. . . . Well, no, I mean it **was** me, a part of me at least, but it wasn't **all** of me, because I'm still here, with or without the arm.*
So what is this "me" you keep referring to? It sounds like you can lop off parts of "it," and yet "it" continues.

ANYONE: *Well, yes, up to a point. If you remove too many parts, though, I wouldn't be here.*
But aren't all "your parts" already being removed on a continuous basis?

ANYONE: *What do you mean?*
You'll recall from our earlier discussion that, nanosecond by nanosecond, countless cells are continuously dying and being flushed from "your body"? And within a few weeks, **all** the material elements that make up "you" right now will be scattered and dispersed into the environment. In just a little more than a day, all the cells of "your pancreas" will be replaced.

ANYONE: *But they're replaced, you see. So I'm still here.*
Even though the material in "your body" is replaced every few weeks, "you" are still here?

ANYONE: *Well, yeah. As I said, I only* **possess** *the body.*
Actually you haven't made it clear whether you think "you" possess "your body" or that "you" **are** "your body." If "you" possess "your body," then "your body" must be something **other** than "you."

ANYONE: *Well, there's also my mind.*
So, in addition to "your body," "you" is also "your mind"?

ANYONE: *Right, I'm not **just** my body. I'm also my mind.*
We haven't yet established whether "you" are "your body" or not, and now you want to drag in "your mind" as well?

ANYONE: *Let's forget all that about the body. Yes, I am my mind. And, as I said, I **possess** a body. That's what I mean: I am my mind and I possess my body.*
"You" is "your mind" and "your mind" possesses "your body"?

ANYONE: *Right.*
Then "your body" is a possession of "your mind"?

ANYONE: *Yes.*
Then "your mind" can exist in some disembodied way without "your body"?

ANYONE: *Well, no. I need my body.*
Why?

ANYONE: *I need my brain.*
Why?

ANYONE: *Because that's where my mind is.*
"Your mind" is in "your brain"?

ANYONE: *Yes.*

How do you know?

ANYONE: *As I told you—as neuroscientists tell us—the brain creates the mind.*

Are you ready to loosen your grip on that?

ANYONE: *I'm . . . I'm not sure.*

Let's come back to that. I need to ask you some other questions first.

13. HOW WE MAKE UP TREES, THE UNIVERSE, AND EVERYTHING

ANYONE: *Can we first go back to when you were talking about perception vs. conception?*

Sure. That's just where I was about to go.

ANYONE: *So when I say, "I see a tree outside my window," I'm supposedly incorrect because, according to you, "tree" is a concept.*

More or less, yes.

ANYONE: *More or less? What do you mean?*

I'm not saying "it's" anything in particular.

ANYONE: *Okay, granted, the **idea** of a tree is a concept, but the tree **itself**—and I mean specifically that tree right outside my window—is not a concept.*

Actually, "the tree itself" **is** just a concept—a thought construct, an abstract idea.

ANYONE: *Abstract? Go over there and rap your head against it and feel how abstract it is!*

There's undeniable perceptual experience. That's not Nothing.

ANYONE: *Right.*

And in this case, we're talking about an undeniable visual and tactile experience.

ANYONE: *Right.*

Is the visual perception of "a tree," though? Or is it of something else? Say color?

ANYONE: *Well, I would say it's of a tree. But, of course, I see color and shape, too.*

Shape is more conceptual—so let's stick with color.

ANYONE: *But that's how I can make out that it's a tree. It stands out against other objects—other colors and shapes.*

So it seems. The concept of "tree" seems constructed from perceptual experience. But the actual visual experience—perception—is simply of light, wouldn't you say?

ANYONE: *Okay.*

Eyes respond to light, not to "trees." By the time we get to "tree," we are deep into concept.

ANYONE: *Maybe so, but that doesn't mean there's no tree.*
I didn't say there's no tree.

ANYONE: *But you're denying that there* is *a tree!*
Not exactly.

ANYONE: *Well, you're certainly not* **confirming** *that there is a tree!*
That's right.

ANYONE: *Come on, you can't have it both ways!*
I'm not saying **either** way. We can't say it is; we can't say it is not. Don't forget what we noted about the law of excluded middle.

ANYONE: *What's that got to do with it?*
When you're caught in the conceptual, the law of excluded middle forces you to grasp what isn't there. In other words, you're allowing a rule of conceptual thought to blind you to direct perception.

All I'm pointing out is that "tree" never appears to the senses. What appears to the senses is light, sound, smell, and so forth.[55]

ANYONE: *Okay. But they all add up to a tree.*

In other words, you're saying that, behind perceptual experience, there's a Real Tree standing there, independent of your concept of "it."

ANYONE: *Yes.*

At first blush, this seems reasonable enough, especially when we consider how our visual impression of "there's a tree over there" is so smoothly corroborated by other sense perceptions that all work in concert with it. It seems that we not only see "the tree," but we can touch "the tree" and smell "the tree." We can even hear "the tree" as wind rushes through "its" branches.

ANYONE: *So, what's the problem?*

The problem is that all this is merely what we believe. Rather than perceptual experience itself, it's our **conceptualization** of perceptual experience.

The sight of "a tree" is a mentally constructed object of sight, not the direct experience of sight itself. And so it goes for all other sense perceptions. Sense perceptions, in other words, are actually objectless.

ANYONE: *But if I'm **looking** at a tree, I'm not making it up. I can't make up an actual tree in the same way that I can make up a thought.*

What, exactly, constitutes an "actual tree"?

ANYONE: *You know, the wood, the bark, the leaves. The tree is **made** of these things.*

Yes—and, in turn, these things are made of other things that you can list. And so it goes on down to atoms, and eventually to electrons and quarks, which, as we've seen, scientists agree are not particular things at all.

ANYONE: *Hmmm.*

What you're calling "a tree" isn't ultimately Real. It's at best a mental report or story **about** perceived Reality.[56]

ANYONE: *Okay. I can see some of what you're saying. Even so, I still don't buy your notion that consciousness isn't generated by the brain.*

If you look at that stubborn belief carefully, you'll see that it comes out of nowhere. We don't even have evidence that consciousness **has** an origin, let alone that the brain produces it.

ANYONE: *Oh, come on! We've got all sorts of empirical evidence that the brain produces consciousness!*

Where? We're a long way from showing how physical processes in the brain can produce subjective experience.

This is the "hard problem" that I mentioned earlier, and neither scientists nor philosophers are anywhere close to an answer. Nor will they ever get to one.[57]

ANYONE: *Why not?*

It's a flat-earth notion, as we'll see.

Mind has no location. Nor is Mind an "it."

14. THE PERSISTENT ILLUSION OF PERSISTENCE

Let's come at this from yet another angle.

We tacitly assume that things exist. That houses exist. That cats exist. That thoughts exist. That feelings exist. That unicorns exist—not as breathing, flesh-and-blood animals, but as ideas and images.

When we say that something exists, we assume and imply that it lasts—that is, that it continues from moment to moment. It persists. It remains itself over time. If it didn't, it would be—or become—something else.

To put it simply, to exist is to persist, and vice versa.[58]

ANYONE: *Mmmm . . . okay. For something to exist, it must remain itself, at least for a time. I'll go along with that.* But persistence of things is an illusion. It's just pure mental construct, an idea.

ANYONE: *So maybe we're more like processes rather than things, like eddies in a stream.*

You're still imagining something that has a continuing separate and distinct existence.

If we look more closely, we can *see* that there's never any actual entity that persists.

We simply imagine we see permanence where there is only change. Yet how can something persist and yet change? How can "it" remain itself and yet become something else at the same time? In fact, what does "it" even refer to under these circumstances?

If things persist as themselves, then they can't change. If they changed, then in what sense are they still themselves?

ANYONE: *Well, I can certainly name some things that persist and yet change!*
Like what?

ANYONE: *Like me, for example. I was once a child, but now I'm grown. I changed. In fact, I'm changing right now. I'm changing in every moment. You pointed that out yourself. Yet I'm still me.*
But in what sense does the word "me" (or "you," or "I") refer to that child? If "me" refers to the body, all the atoms of that child's body have long ago dispersed into the environment. If "me" refers to the mind, then every thought, feeling, or mental impression experienced by that child has long ago vanished.

ANYONE: *Yeah, but I have memories, photos, and even videos and recordings of me as a child. I can even remember when many of them were made. I was* **there***!*

Sure, you have memories and photos of that child, but "the child" no longer persists. Nor did "it" ever. There's just the immediate "me"—the "you" right now—who is obviously not the same as the vanished child.

ANYONE: *But I have the actual memories of that child!* **They** *persist.*

The memories you speak of are not frozen in the past. They're *now*. You remember them *now*.

Furthermore, whatever memories you experience *now* are not playbacks of mental videos of "then." Anyone who has studied human memory can tell you this. You're creating them *now*, not retrieving them fully formed from some neurological archive.

Nor will your memories remain as they are *now*. Memories continuously mutate. In fact, every time you "recall" a memory, you alter it.

In any case, the past simply isn't *here*. Nor is the future. There is **only** *now*, memories included.[59]

ANYONE: *Okay. The child is past and the past is gone. And the future is not here, either. And memories mutate. I grant you all that. But I can say that that child* **was** *me, even if it's not me now.*

That brings us back to my earlier question: How can something be itself and yet change? The instant something changes, "it" is no longer itself, but something else.

There's only the **appearance** of being—the appearance of existence, of permanence, of persistence—not the actuality of it. In a very limited sense, "you" only **appear**, as a mental construct, in some rather vague sense as being "the same person" of a year ago, or twenty years ago.

The "you" of one year ago, or one week ago, or even one moment ago, is not the "you" of *now*. "You" never persisted; this imagined "you" seems to be continually changing. In fact, there's no "you" "there" **to** change.

There **never** was a persisting "something" to which the word "you" could apply. Nor is there one *now*.

"You" doesn't refer to anything Real. And the same is True of any entity we can imagine or conceive of.

"You," "me," "I," "it," or "cat"—these terms don't refer to some Real Entity, but only to the appearances of thoroughgoing change itself.

All we ever find is change.

We can *see* this if we *attend* closely. We can't find a thing that endures, even for a moment—even for a nanosecond.

ANYONE: *So, you're saying that everything is impermanent and ever changing.*

No. You're positing "things" that have the quality of impermanence. You're invoking two concepts, as in "fat cat"—things and impermanence—and then layering one on top of the other.

I'm not saying that everything is impermanent. I'm saying we don't actually find permanence anywhere—and that we don't find thingness, either. There can be no things that are impermanent, since "their" very thingness would be contingent on persistence.

Impermanence is thorough. So thorough, in fact, that there are no things to **be** impermanent in the first place.

ANYONE: *Then why can I see them?*
What you actually *see* is thoroughgoing change.[60]

What you **think** you see is permanence, thingness. But that's merely an appearance, a concept, an illusion. We directly *see* Reality; then we assume thingness, persistence, and existence.

ANYONE: *But I wasn't really assuming anything.*
No? Do you assume this floor will hold you if you take a step?

ANYONE: *Sure.*
Do you assume it's safe to take another breath?

ANYONE: *You're saying our heads are full of unseen, unex-amined, unchallenged assumptions.*

We hold lots of such assumptions, yes. Whether those assumptions reside in our heads is another matter.

15. FORGET WHAT HAPPENS

Let's look carefully at what, for now at least, we can call our *objects of consciousness.* These are the "somethings"—the concepts—we've been imagining and talking about.

Before we dig into this, though, I need to clarify just what I will refer to as *consciousness.*

Consciousness is discrimination.[61]

ANYONE: *I thought you would say "awareness," or "sentience," or something like that. Could you please explain what you're getting at?*

For the most part, we deeply believe that "things" are just "out there" and that "our consciousness" "takes them in."

As I noted earlier in several examples, particularly with Virgil regaining his sight (see the *Time Out!* on page 66), any object of consciousness—the "thing, out there"—is pure mental construction.

So, let's not start with the deep-seated flat-earth belief that there actually **is** an "objective world" "out there," filled with all sorts of "things" that come and go, and that "my consciousness" "takes them in." If you really want to understand conscious experience—and thereby, consciousness—you need to realize that all such "entities" are no more than manifestations of Mind.

Thus, we can directly *see* the illusion. More than that, if we stop clinging so tightly to our belief that the illusion is Real, we can begin to *see* the delusion. We can *wake up* to the fact that it's just an idea that "there are" "things" "out there" that "I" "take in" "through" "my senses."

In order to truly accomplish this, however, you need to set aside all such beliefs and *attend* to actual experience.

There **does** appear a "this" as opposed to "that." There appears this color as opposed to that color. And, on a broader scale, there appears this sound, as distinguishable from that heat. There appears a variegated field of many impressions and sensations.

Yet we can avoid ascribing any objectivity or substance to what appears. We can simply take note of the distinctions among the appearances of "this" and "that."

ANYONE: *Okay. And then what? What does all of this add up to?*

These apparent distinctions constitute what we can call conscious experience.

Or, to put it another way, consciousness is the marking of distinctions that give rise to the illusion of thingness, and the discernment of mentally constructed boundaries.

ANYONE: *Hmm. And how do these boundaries arise?*
With the appearance of things "out there," there appears the parallel sense of "me, here," taking in those appearances.

This is the Grand Illusion. There appears a world, not just "out there," filled with objects and beings, but also "in here," filled with feelings, thoughts, ideas, and other impressions.

Thus, the Grand Illusion parallels our collective Grand Delusion: our deep, unquestioned conviction that such phenomenal appearances are True and Real, rather than persuasive (as well as pervasive) illusory distinctions.

And all of this appears to play out within time and space. But, as we'll see, time and space are illusions, too.

All such distinctions are no more than appearances. In our day-to-day world, they're often functional and useful, and need to be given their due. But we need not mistake them for Reality. In fact, we make this mistake at our peril.

ANYONE: *So, what is the reality, then, behind these appearances?*

You're still imagining a substantial "Something" behind appearances. You're also imagining a "behind."

There's no particular Something behind the appearance of "this" and "that." There can't be any supra-substantiality behind nonsubstantiality, some Something behind the reality of thoroughgoing change. There's also no "behind"—no curtain to peel back, either.[62]

It's quite possible to *just perceive* This—to be fully *aware* of experience without putting a word, or a story, or a thought, or an explanation to any of it.[63]

Reality can be *seen* and *known* directly. Indeed, it can **only** be *seen* and *known* directly.[64]

Words and concepts will never capture It. Yet words and concepts are what we reach for every time, while ignoring the immediate and True.

ANYONE: *So, what are we supposed to reach for instead?*
There **is** no instead. Simply stop reaching. Just *pay attention*.

ANYONE: *To what, though?*
Forget "what." Just *pay attention*. Not to anything in particular.

Either we're paying attention or we're not. And paying attention does not involve reaching, or words. Or concepts.

ANYONE: *So, I should . . . ?*
Just see.

Don't add anything—thoughts, words, names, labels, commentary—to what is directly experienced. Instead of endlessly reaching for ideas, for explanations, for concepts, *just see*—and let it go at that.

ANYONE: *And then what happens?*
Again, you're grasping. Forget "what happens." Forget "then." Forget "and." Just let it all go.

16. DIFFERENT FROM ANYTHING ELSE

ANYONE: *What about all the things we* **can't** *label, or grasp, or discuss, or name, or imagine?*

Such as?

ANYONE: *Well . . . I don't know. There must be all sorts of such things.*

Consider what you're saying. You're imagining that we can call to mind things that can't be called to mind. If you are interested in *awakening* to actual Truth and Reality, it's imperative that you stop clinging to any and all beliefs, concepts, words, assumptions, and views. Truth never forms as an object of consciousness. It will never go into a concept—and It can never be expressed as one, either.

ANYONE: *And how do I stop clinging, exactly?*

You simply stay with immediate, direct experience. You stop putting everything into words. You stop talking to yourself about what any particular experience is, or means, or how true it is, and so forth.

The more you see how strangely Nature behaves, the harder it is to make a model that explains how even the simplest phenomena actually work. So, theoretical physics has given up on that.[65]
—*Richard Feynman*

ANYONE: *Is that really possible? It sounds difficult.*
With some effort, concentration, and honesty, it can be done; you're just not used to making the effort.

ANYONE: *I never thought I had to!*
Right. But if you're interested in finding out where you're confused, that's where you need to start. And it will be uniquely different from anything else you might do. But it's quite possible. And actually, quite simple.

ANYONE: *Okay. So, I . . . ?*
I recommend meditation. I wrote a book titled *Meditation Now or Never* that will help you in this regard.

ANYONE: *Meditation?*
Yes, but of the proper sort. There's a lot of stuff out there that passes for meditation—trance-inducing exercises, relaxation techniques, and whatnot. They may claim to

be meditation, but they're not what I'm talking about. In fact, they're hindrances to staying with immediate, direct experience.

ANYONE: *You're talking about mindfulness.*
The word "mindfulness" has gotten corrupted and commodified, so we'd be wise to avoid it. I'm speaking of Awareness. We need to *see* Reality directly without wandering off or framing It in concepts.

If you're interested in *waking* to Truth and Reality, you need to confront The Grand Delusion—the profound but false conviction that there actually **is** Something rather than Nothing.

17. TASTING ACTUAL KNOWLEDGE

True Knowledge transcends concepts. It's directly and wordlessly experienced. In other words, True Knowledge is objectless, as we shall *see*.

ANYONE: *But we're often mistaken about experience.*
Conceptualized experience, yes. But when, say, you mistake the sound of distant thunder for a plane, you're mistaken about what you **think** you experience, not about actual experience itself. We're not actually confused about perceptual experience at all—at least if we don't think about it. If there is any confusion, it comes only when swapping one concept for another—e.g., thunder for a plane—or when taking concepts for Reality.

In other words, we can't be mistaken about direct experience. We can only be mistaken about mentally formed **objects** of consciousness—about our conceptualized **interpretation** of perceptual experience.

ANYONE: *Hang on. What if you saw flashes of light, but only experienced them because you just got hit on the head? You saw a flash, and you experienced lights, but they were nothing more than electrical impulses in your brain.*

What you're calling a flash is nevertheless directly experienced. The only mistake is in the conceptualized interpretation: "I just saw lights flash on."

ANYONE: *Then is everything we experience nothing more than electrical impulses in our brains?*

I wouldn't get too carried away with the idea of the brain as the seat of sight or sound.

ANYONE: *Why not? I mean, that's where sensations occur. Say it's not simply flashes from being hit on the head. Say you're asleep in bed, dreaming. There are electrical impulses flashing in the visual cortex of the brain, just as if you were seeing with your eyes.*

As we shall *see*, perceptual experience isn't located anywhere.

ANYONE: *What do you mean?*

Before conceptualizing "it," perceptual experience is unformed and unlocatable. Yet it is what we can rely on.

Through wordlessly scrutinizing direct experience—through *seeing* what is going on before making anything of it—we can realize True Knowledge.[66]

ANYONE: *You mean knowledge that's meaningful.*
No. Think about what you're saying. Meaning puts us back into the conceptual again.

We regularly confuse meaning with actual Knowledge. But meaning is **always** conceptual. When we say "this" means "that," one word or concept stands in for another.

True Knowledge—direct perception—can't mean something else, or be **like** something else. It's immediate.

True Knowledge—the taste of orange juice, for example—is immediately and wordlessly *known*, and that direct experience is not conceptual. That taste—that perception—is an example of True Knowledge.

Any description of that perception, however, is an abstraction. Even naming it "orange juice" is an abstraction, a concept added after the immediate experience (or inserted in place of it).

Concepts never supply True Knowledge.

ANYONE: *Then give me a taste of true knowledge.*
I can't give you what you already *know*. I can only point out where and what you're habitually ignoring.

ANYONE: *Hmm. Then continue, please.*

18. SLOW DOWN

All ideas about consciousness are wrong.

You need to *see* this directly.

I'm not interested in giving you **ideas** about consciousness. Rather, I want to help you drop the ideas you already have.

If you don't let go of whatever you think you know regarding consciousness, you'll not *wake up*.

You need to let go of all the ideas that are commonly accepted these days: that consciousness has an origin; that it's produced by the brain; that it's **my** consciousness or **your** consciousness; that it's **my** cognitive take on a world "out there"; that it's a given that there **is** "a world out there"; that the word "my" actually refers to what is Real; that consciousness is what makes me human (as opposed to animal); that it's what I have language for; that it's my self-awareness; and so on. Whatever they are, drop them all.

ANYONE: *So, what do I do instead?*
Simply *pay attention*.

Now slow down.

DON'T ENGAGE THOUGHT.

Start by wordlessly taking note that This is not Nothing. Also note that This is immediately *known*.

STOP!

Take a moment.

Now notice: there may (or may not) appear a visual field. If so, it appears ever-moving and variegated—light colors, dark colors, shapes, sizes, visual textures.[67]
There, too, may appear a field of sound. It appears variegated as well—high-pitched sounds, low-pitched sounds, sounds fading in and out, some lacking any intonation, some harsh and sharp, others round and soft.
There possibly appears, strengthening and weakening, a field of smells. It too appears, for lack of a better word, variegated.
There may appear a field of flavor and taste. Also variegated.

Like so, a variegated tactile field may appear. And, just short of thought, an undulating variegated field of emotions appears.

There appear other, subtler, shifting fields as well, plus varying depths of time and space. And all appear variegated, mottled, dappled in endless variety, in unending difference.

This color differs from that color. But also, this color is distinguishable from this aroma, this pain, this intonation, this thought. Distinctions of all sorts appear.

What does **not** appear, prior to any analysis or synthesis, is substance—some thing or things—behind the endless variations.

Ultimately, you have to *just see This.*

Time Out!

Put this book down. Just sit quietly and *pay attention.*

Lightly and wordlessly attending to breathing may help at this point. Don't worry if thoughts arise; just don't pursue them. Just return to the breath when you notice that your attention has wandered away. No need to say anything to yourself about having wandered away. No need to comment on anything. Keep it wordless as best you can. Just stay present. **Pay** *attention.*

Continue this for a while. No more than thirty minutes. (We will soon see how time is an illusion. Even so, as little as five minutes is more than adequate.)

19. CONSCIOUSNESS, AWARENESS, AND REALITY

Without presuming that actual objects exist (or that they do not exist), the world appears in every moment of conscious experience to bloom with endless variety. Yet immediate experience itself is entirely without graspable separations and boundaries.

ANYONE: *Huh? Isn't that a contradiction?*
Only if you try to grasp It conceptually.

Look for a boundary. It would seem that one ought to be there whenever you experience differentiation. But you'll never actually find one. Any boundary is just an idea you're holding—an implicit belief, a concept, a mental construct. It's not Real.

ANYONE: *Can you give an example?*
Not of anything in particular. *Just look.*

ANYONE: *Look at what, though?*

Even specifying a "what" misses the mark. But look at the shifting hues of a sunset. The change in colors and the quality of the light never stops. Even so, if you attend closely, you'll never actually see one color become another.

Or, closely attend to any irregularly recurring change—the predawn song of the robin, the late-night hoots of an owl, or even the slow, irregular drips from a leaky faucet. If you let your mind go quiet, if you don't think and just *wordlessly* attend, the sound either sounds or it does not—but if you attend carefully, you'll never actually hear "it" starting or stopping.[68]

If you attend carefully, you can *see* that this is conscious experience: the **apparent** dividing of what otherwise remains a seamless Whole.[69]

ANYONE: *So, what are you saying?*

There are two aspects of Mind: consciousness, the apparent dividing up of Reality; and Awareness, or undivided Wholeness.[70] The former is conception; the latter, perception.

If we look carefully and honestly, we can *see* that conscious experience always involves an object, as in "I see the bird outside my window" or "I felt the pain in my knee." And it also always involves a subject: "**I** see the bird." "**I** feel the pain."[71]

In contrast, pure Awareness has neither subject nor object—nor any need for them. There is simply perception—and Wholeness.

20. THE SELF ILLUSION

Consider a lake. What is it? What makes "it" a lake?

Rain falls in; water evaporates from it and seeps into the soil beneath; multiple streams flow in; a single stream flows out. The ever-changing flora and fauna of the lake—are they parts of the lake? If so, what about the waterfowl that come and go? Or does the lake consist only of the water that surrounds their bodies? If so, is that also true of the fish that take in water through their gills? The water flowing in and out of those gills?

A "lake" is a mentally conceived object. There's no Real Lake that can be pinned down or gotten hold of.

A lake is a concept.

The same can be said of a cat, a cloud, a star, you, God . . . anything.

Yet we also cannot legitimately say there is no lake, or cat, or cloud, and so on.

Whatever we consider, it is neither Nothing nor Something.

ANYONE: *Okay, so a lake is an ecosystem.*
That's just another, seemingly larger concept you're grasping at.

We can go in the other direction, too. Are you aware that there are more microbes living in your mouth right now than the total number of humans who have ever lived on Earth?

ANYONE: *I did not.*
Did you know that there are 100 trillion cells in your body, and of those 100 trillion cells, only 10 trillion are what we might loosely define as "you"?

ANYONE: *What do you mean?*
Ninety trillion of those cells are actually bacteria. Only one in every ten cells in your body carries "your" DNA.[72]

ANYONE: *There you go! DNA! That's what defines us.*
Yes, but you also need those ninety trillion bacteria in order to be "you." Your gut and maw, for example, as well as the other organs in your body, couldn't function without them. "You" couldn't live without them. In fact, there'd be no "you" to speak of without them. They're essential to you being you, whatever "you" is supposed to mean, given that "you" is as much those "other" ninety trillion cells.

In short, "you" or "I" never refers to anyone or anything in particular.

ANYONE: *But there's still* **someone**.
Can't say there is. Nouns and pronouns **never** refer to anything in particular—not in Reality. Only in concept.

ANYONE: *Can you give an example?*
Have you ever used the phrase "it's raining"?

ANYONE: *Of course.*
What do you suppose is raining?

ANYONE: *Pardon me?*
What do you think the word "it" refers to?

ANYONE: *Ah . . .*
That's one case where we commonly refer to something—ostensibly, at least—even though no one actually thinks there really **is** something that's doing the raining.[73]
Consider the phrase, "I'll meet you at sunrise." Does that phrase strike you as meaningful?

ANYONE: *Yes.*
But the sun doesn't rise.

ANYONE: *Pardon?*

The sun doesn't actually rise. It doesn't actually propel itself upward from beneath the horizon and fly across the sky, as ancient people who coined the phrase believed.

ANYONE: *Yeah. Of course. So what?*

We know better.

ANYONE: *Well, yeah. But . . . .*

It's just an appearance.

ANYONE: *Yeah.*

But we continue using the term *sunrise*, since that's a lot easier than saying, "Let's meet at that moment in the day's diurnal cycle when, due to the Earth's rotation, the first glints of sunlight burst forth on the eastern horizon," or words to that effect, right?

ANYONE: *You took the words right out of my mouth.*

So, you would agree that it's convenient and useful to make use of that reference—to sunrise—even though we know it refers only to an appearance rather than to an actual event?

ANYONE: *But it* **does** *refer to an actual event!*

The first glints of the sun, you mean.

ANYONE: *Yes. Unlike your dismissal of you or me as mere appearances, which is what I think you're implying, it* **is,** *nevertheless, an actual event beyond mere appearance.*

There you go again with that word.

ANYONE: *What word?*

"It."

ANYONE: *Yeah. So?*

Convenient as it is, what does "it" refer to? What is the "actual event"?

ANYONE: *Well . . . .*

"Sunrise" is a thought construct. So are "you," "me," and "I." Thought constructs make it easy for me to ask you to pass the salt. But these thought constructs— **all** thought constructs—do not refer to anything Real. They're quite useful, but they only refer to what **appears** to be going on.

21. A SELF WOULD HAVE TO BE SOMETHING

Would you agree that for a thing to be what it is, it can't be something else?

ANYONE: *Well . . . yeah.*
Notice that this never occurs.

ANYONE: *How does **that** follow?*
As we've just noted, nothing is anything in particular. Everything changes—or, at least, so "it" appears.

ANYONE: *I don't follow you.*
A self would have to be something that **doesn't** change.

ANYONE: *Why?*
If it changed—if it became something else, in other words—how could it still be itself?

ANYONE: *Okay. I'm with you.*

Therefore, by definition, a self cannot change. A self, in other words, would necessarily have to persist **as** itself even in the midst of thoroughgoing change.[74]

ANYONE: *Okay.*

But we've already seen that the world is nothing **but** change.

ANYONE: *Hmm.*

There's a subtle but profound contradiction here. That's why you're having a hard time *seeing* the nature of Reality.[75]

Simply put, you believe that mind objects are not merely existent, but primary.

ANYONE: *What do you mean, "primary"?*

You believe they precede Mind.

ANYONE: *Don't they?*

Not if they're manifestations of Mind.[76]

There are no examples of objects preceding Mind. Nor could there be.

ANYONE: *So, mind is primary, and . . . ?*

And Mind is nonsubstantial.

22. THIS ILLUSORY WORLD

We live in a seemingly substantial world. But, as we've seen, on close examination it loses all substantiality (not that "it" ever had any to lose). Indeed, it's unpindownable.[77]

Yet throughout that examination—and that nonsubstantiality—Mind remains ever present.

Mind is the basis for the material world, not the reverse.

ANYONE: *So, you're saying that my mind . . . ?*
Who said anything about **your** mind? Stop placing yourself in the middle of everything. (Not that everything has a middle, or an edge, or an end. Or even thingness.)

You still imagine that Awareness, direct experience, somehow occupies physical space—that consciousness resides in matter, or is created by it.

ANYONE: *But isn't it?*
Does that in any way make sense to you? That consciousness has a location in physical space?

Look at the book you're holding. You see the book; you feel it; you can hear it if you rap on it.

ANYONE: *Right. So?*
Where is that experience taking place?

ANYONE: *In my head, of course!*
The book is in your head?

ANYONE: *No! The* **experience** *of the book is in my head! That's what you asked about . . . the experience. It's in my head.*
But the experience is of something presumably outside your head, wouldn't you say?

ANYONE: *Of course!*
So how can you say it's in your head? Your head and the book seem to be in two different places.

ANYONE: *So, you're saying the experience is outside my head?*
I'm saying that experience doesn't seem to have any location at all.

ANYONE: *You're kidding!*
This is an obvious absurdity we run into whenever we try to reduce everything to the physical. Where is anything occurring? Sound or smell or sight? Even taste or touch

or thought. Only the physical components—the physical **objects**—of experience seem to have location, not the immediate experiences themselves.

Yet remove any component of the experience, no matter its ostensible "location," and there **is** no experience.

ANYONE: *What do you mean?*

You hear a bell, smell a rose, see a bird. Where does seeing, hearing, smelling happen? In the bell or rose or bird? In the sense organ that "picks it up"? In the space in between? In the neurons of your brain? Remove any bit of this and there's no experience. So, where is it happening?

This is important to notice: conscious experience **has** no location. It's pure mental phenomena.

You can *see* this directly. Just turn your attention to it. It's immediately obvious.

Simply put, **Mind** is the basis for the so-called material world, not the reverse. And the material world isn't actually material, but completely nonsubstantial.

23. THIS SIZELESS WORLD

There is much else that points to the primacy of Mind over matter.

ANYONE: *Like what?*

If you try to move in on Mind for a closer look, as we did with matter in chapter 7, it doesn't make sense. There **is** no "moving in." Since Mind has no size, It doesn't expand, shrink, or disappear.[78]

Your mind might give these appearances, but "your mind," or "my mind," merely refers to thought objects, concepts, ideas, notions, and figures of speech.

We're not talking about something relative, here. We're talking about Reality. Mind. Totality. It's not an object. It has no location and It has no size.

Thus, the Universe, *seen* as Totality, is neither big nor small. It, too, has no size.

ANYONE: *The universe has no size? I gotta hear this!*
Well, look at how oddly It presents Itself when you assume It's objectively real.

ANYONE: *Like what?*
Consider the light from a star that's a thousand light-years away.

ANYONE: *What about it?*
We say that the light from that star took a thousand years to reach us, traveling at 186,000 miles per second. That's nearly six quadrillion miles, in fact.

ANYONE: *Really?*
Kind of a questionable figure, though, wouldn't you say?

ANYONE: *Why do you say that?*
It's incomprehensible.[79] More than that, it's ultimately meaningless.

ANYONE: *How so?*
Consider a photon arriving from that star. If you could read the photon's clock while it was in transit, it would appear to you that, for the photon, it took no time at all to get here. Yet, according to your own clock, it would seem to have taken a thousand years for the photon to arrive.[80]

ANYONE: *Why?*

Photons travel at the speed of light, and at lightspeed time doesn't elapse.

ANYONE: *Oh. Yeah. Einstein and all that. But we were actually talking about size—you know, space, not time.*

Well, traveling at the speed of light, the "space traversed" by the photon is zero—if **you** could measure it using the photon's superfluous yardstick. According to **your** yardstick, however, the photon traveled nearly six quadrillion miles. You know, Einstein and all that.[81] So, as far as **your** view of **the photon's** reality is concerned, it didn't cover any space or distance at all. And it didn't take any time. From **your** perspective, the photon would seem to experience no space, no time, no location, no motion.[82]

ANYONE: *Okay, but we're not photons.*

The question is, what gives your frame of reference privilege over that of the photon? Or, more to the point, over Totality?

What I'm inviting you to contemplate is that, ultimately, the Universe, in Totality, is neither big nor small. It's intrinsically without size. You need to let this sink in.

ANYONE: *But that makes no sense!*

Be careful. Much would be clarified immediately if you'd simply allow that it's the way we frame things in our minds that ultimately makes no sense. When we insist that our conceptions are ultimately Real, that's when we create contradictions and mysteries and confusion.

In other words, what we call "the universe" is merely a thought object—a mental construct, not an objectively Real Thing.

ANYONE: *So you're saying it's* **not** *really out there?*
Absolutely not. I'm only noting that we can't say that it is.

ANYONE: *Oh, come on!*
For that matter, we can't say there's even an "out there," given the implications of Bell's theorem.

ANYONE: *What's that?*
It requires Reality to be non-local.

ANYONE: *What does* **that** *mean?*
In a local reality—that is, in a reality where "here" appears to be in a different location than "there"—influences cannot travel faster than the speed of light. In 1964, however, using no more than simple arithmetic, mathematical physicist John Stewart Bell was able to show that, given such a commonsense view—i.e., that "here" differs from "there"—information cannot get around the universe fast enough to account for the quantum facts.

ANYONE: *Meaning?*

There's no here vs. there. There's just *here.*

> It's neither long nor short, big nor small, for it transcends all
> limits, measures, names, traces, and comparisons.
> —*Huang-Po*[83]

ANYONE: *Hmmm.*

There's a great deal of experimental evidence that backs
this up, and virtually none against it.[84]

ANYONE: *Okay. So, what are you getting at, exactly?*

Just as Mind is without location, size, or substance, what
we call the Universe is likewise without location, size, or
substance.

ANYONE: *Which means . . . ?*

The World is Mind.

> There is only the One Mind and not a particle
> of anything else on which to lay hold. . . .
> —*Huang-Po*

24. MIND IS MOVING

Just as thingness, space, and time are pure mental constructs, so, too, are matter, energy, and motion.

ANYONE: *How so?*
Smash two particles together—say, two protons—at extremely high speeds, as in a particle collider. What do you suppose happens?

ANYONE: *I think you end up with more matter, don't you?*
Right. The original particles fly apart, along with **additional** particles that appeared nowhere in time or space "before" the collision.

This is nothing new. Such observations have been noted for years. Quite routine.

But where did these new bits of matter come from? Physics tells us that they formed from the **reduction in speed** of the original two colliding particles.[85]

So, how substantial **is** matter, then—this page, this hand, this scanning eye, this flickering brain—if it can all be created from something as nonsubstantial as motion?

Despite recent assertions to the contrary, modern philosophy and mathematics have not been able to put away all of the inherent contradictions of motion discovered by the Eleatics.

ANYONE: *Who?*

The ancient Greek philosophers—among them, Zeno—who noted, more than twenty centuries ago, that a thing can move neither where it is nor where it is not.[86]

Time Out!

An unsuccessful attempt to set this paradox of motion to rest was made by William I. McLaughlin in his article, "Resolving Zeno's Paradoxes," in *Scientific American*, November 1994 (reprinted in 2006).

As the basis of his argument, McLaughlin posits the ostensible reality of "infinitesimals."[87] McLaughlin defines such entities as "greater than zero but less than any number, however small, you could ever **conceive** [my emphasis] of writing." He then concludes that "an infinitesimal interval can never be captured through measurement; infinitesimals remain forever beyond the range of observation." In other words, they can neither be perceived **nor** conceived.

Motion, says McLaughlin, somehow resides in these infinitesimals, and with that fanciful notion, according to McLaughlin, all of Zeno's paradoxes of motion are finally put away.

Unfortunately, since these infinitesimals can neither be conceived **nor** perceived, they are neither Real nor imaginary. They belong to neither the realm of matter and energy nor the realm of thought and imagination. This is hardly a resolution of anything, but merely a chimera masquerading as a proof.

Another unsuccessful attempt—a rather unconventional one, but not that dissimilar from McLaughlin's—was put forth by Peter Lynds, who also claims to do away with all of Zeno's paradoxes. Lynds feels that he accomplishes this by simply declaring the absence of any instant in time in regard to the motion of a body by dismissing it as a phantasm of consciousness. According to Lynds, "... no matter how small the time interval, or how slowly an object moves during that interval, it is still in motion and its position is constantly changing, so it can't have a determined relative position at any time, whether during a[n] interval, however small, or at an instant. Indeed, if it did, it couldn't be in motion."

Spoken like a true substantialist. And, perhaps, it sounds good at first blush, except for the "it is still in motion" part, which appears to have been pulled out of thin air. Once again, the error lies in the unquestioned assumption of substantiality or particularness—i.e., in Lynds' assumption of an "it"—and, beyond that, in the assumption of motion as an objective, material reality as well.

Be that as it may, Lynds **is** correct in linking the instant—effectively, the quantum, the packet, the concept—to consciousness. Unfortunately, he errs in dismissing the instant on those grounds. In effect, what Lynds is saying is not all that different from McLaughlin. Both assume, as substantialists **must** assume, that motion just is. It never occurs to either of them that they can't legitimately say "is," and can't legitimately say "is not." They do not see that motion is no more than an appearance resulting from the functioning of consciousness.

Of course, this observation only poses a problem for those who hold to a substantialist view, believing that stuff is really "out there" and can actually be pinned down—or moved around.

ANYONE: *Huh?*
How can a thing be where it isn't? Or possibly move to where it already is?

Of course, as we discussed, we can't rightly say that there **is** a thing—an "it"—in the first place. Nor is there a "there"—a place for "it" to be located in, let alone to move from or to.

ANYONE: *No "there"?*
Don't forget what Bell showed us about non-locality.

ANYONE: *Well, then, if there is no "there" there, then how can we have motion?*
Can't rightly say that we do.

ANYONE: *Of course we do! It's obvious. How could we not?*
Only in appearance, not in Reality. Upon close investigation, we discover that motion is illusory. Purely Mind.

ANYONE: *Come on!*
We've already noted the illusory nature of time and space.

> **Not flag. Not breeze. Mind is moving.**
> **—*Hui-Neng*[88]**

ANYONE: *Remind me.*
From our perspective, there would seem to be no space or time for the photon. Ultimately, this means we can't legitimately say that there's location or motion, either.

Just like matter, space, and time, location and motion are simply constructs of Mind.

ANYONE: *So, let me get this straight. You're saying* . . .
Motion is conceptual. It's not ultimately Real.

ANYONE: *Surely you can't be saying that there's no motion.*
No. Of course not. We can't say that there's no motion.
It's just that we also can't say that there is.

25. HOW MOTION IS MIND

Let's try a little thought experiment. Consider the Koch curve.

ANYONE: *What's that?*

Picture a straight, horizontal line drawn from point A to point Z, as shown immediately below.

A   Z

Let's divide that line into three equal segments—AH, HS, and SZ—as illustrated:

A   H   S   Z

Now, pivot the middle segment, HS, 60° counterclockwise from point H, as shown here.

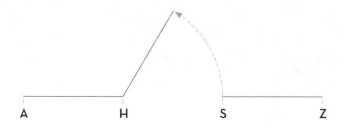

ANYONE: *Got it.*

Pivot a copy of that same middle segment, HS, 60° clock-wise from point S (see below). Note that the opposite ends of these duplicate middle segments meet at point M, which also marks the midpoint along the new, zig-zagged AZ line. Note, too, that segments HM and MS form two sides of an equilateral triangle, the third side of which (i.e., the space between points H and S) is now missing.

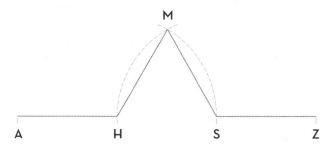

Notice that we have now increased the length of the original line, A to Z, by one third (we've multiplied the line's length by ⅘, in other words).[89]

ANYONE: *Okay, I can see that. We originally had three segments of equal length, marking a straight line from A to H to S to Z. Now we have four segments of the same length running along the zigzagged line from A to H to M to S to Z.*

We can repeat this process by doubling the middle third of each these four new segments and pivoting them in this same zigzagged manner, thus increasing the length of the line from A to Z once again by ⁴⁄₃.

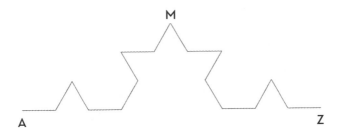

ANYONE: *Okay. I got that. Each segment between A and Z has been increased in the same manner as before.*

We can continue this process of increasing the length of line AZ through a limited number of such iterations. Note that, as we do this, the distance between points A and Z, "as the crow flies," does not increase.

ANYONE: *All right. I'm still with you.*

So, if you moved a cursor along this ever-increasing and ever-more-meandering AZ line, starting at point A, how long would it take the cursor to arrive at point M?

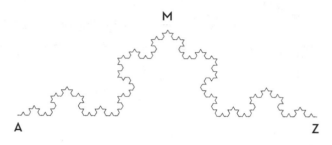

ANYONE: *That would depend on the exact length of the original straight line between points A and Z, how many iterations we've made, and how fast the cursor is moving.*
Right. But it would eventually get to point M?

ANYONE: *Of course.*
But what if we continued this iteration process through an infinite number of steps, thus creating a so-called Koch curve—a line of infinite length running from point A to point Z?

ANYONE: *Ahh . . . wait a minute.*
Well, starting with a length of 3 (counting 1 for each of the three segments on the original straight line running from A to Z) and then multiplying by ⁴⁄₃ with each iteration, we now have 3 × ⁴⁄₃ × ⁴⁄₃ × ⁴⁄₃ . . . . And so, after an infinite number of such iterations, the line's length stretches to infinity, right? [90]

ANYONE: *Okay, I'm still with you. In fact, I can see that, iteration after iteration, the length of the line increases at an accelerating rate, since each iteration multiplies an ever-lengthening line.*

Yes. Now, after an infinite number of such iterations, starting at point A, how long will it take a cursor to move—remember, we're interested in understanding motion in this chapter—along the zigzagged line to reach point M?

ANYONE: *Ahh . . .*

Say we situate ourselves near point M and watch and wait for the cursor to arrive at point M. How long will we wait? A minute? A year? Ten thousand years? Ten billion years?

ANYONE: *Hmmm. . . . It will never arrive.*

Right. So where will the cursor be if, say, we check on it after a billion years, with the cursor clipping along all the while at some ultra-high speed along these infinitesimal zigs and zags?

ANYONE: *Hmm. I'm not sure.*

Suppose we place a circle around our cursor at the starting point (point A), so that we can track it as it clips along. (An infinitely small cursor would, of course, be too tiny to picture. But this doesn't keep us from tracking its location by positioning it at the center of a visible circle.)

Take as much time as you want and make the cursor "move" as fast as you want—but you will never see the

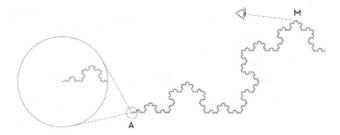

the tracking circle move. Movement is not to be found within infinitesimally small units of length or time. If it were, there'd be no way to fully understand consciousness—as our popular, current-day ideas about consciousness amply demonstrate.

You've got to have **finite** (i.e., conceptualized, or quantized) lengths in order for the cursor (or any object) to render any appearance of motion. Movement is in the appearance of movement. It's strictly a phenomenon of Mind.

The insurmountable problem for the substantialist is that we can't conceptualize infinities (including infinitesimals) as objects. Thus, only so long as the zigzagging AZ line remains finite (i.e., remains a conceptualized object) will the cursor appear to move. The moment the line becomes constructed of infinitesimally small segments, the cursor will cease to give the appearance of motion.

Motion—like all phenomenal entities—is pure mental construction. It's conceptual rather than perceived Reality, as we naïvely take it to be.

The fact is, the World has to—let me repeat: **has to**—appear this way. It has to **appear** quantized, as though it comes in finite packets, exceedingly small though they may seem. Either this or there'd be no appearance of anything at all—including motion.

There'd **be** no conscious awareness, in other words. This is, after all, what consciousness is.[91]

As for big R Reality—which is available to perception but not to conception—when it comes to motion, time, space, or any other conceptualized entity, we can say neither "is" nor "is not."

Which is precisely what we found with point particles; with the size and age of the Universe; with ourselves; and with what we will find with any number of otherwise inexplicable phenomena.

All are fabrications of Mind.[92]

ANYONE: *Are you then saying that perception is unconscious experience?*

There's no such thing. I'm only pointing out that pure perceptual experience—that is, experience without conception—is objectless.

ANYONE: *So, you meant it literally, then, when you said that the world is mind?*

Indeed. Mind is the World.

26. THERE IS ONLY MIND

As we look at the World more and more closely, all reduces to Mind. Or, more accurately, there's no reducing. All **is** Mind.[93]

> When I point out that Mind is grasses and trees . . .
> it startles your ears.
> —*Dogen Zenji*

Yet almost everyone—scientists, philosophers, theologians, butchers, bakers, and microchip-makers—insists that everything can be accounted for by matter alone. That even the immediate experience of matter itself is no more than what takes place in the electrical circuitry of a person's brain.

But how do we get from chemical-electrical circuits to, say, awareness of smell? We never even approach an answer to this question. Instead we take a giant leap of faith—and in the process, wall ourselves off from any kind of understanding.

ANYONE: *But you've got to admit that my brain is involved.*
So it seems. We've also noted that your eyes are involved, and your nose is involved, and your ears, and the air, and the light—and, indeed, the rest of the universe.

We've also noted that matter is *empty* of any substantiality. We've noted, too, that space doesn't have any intrinsic size, and that "here" and "there" are not intrinsically distinct.

We've noted all this, yet mental phenomena take place.

ANYONE: *So, you're saying it's all in my mind.*
We've also noted that words like "I," "you," "me," and "my" don't refer to anything in particular.

ANYONE: *Then who is seeing and hearing?*
If we note carefully, it's all *just* Mind.

ANYONE: **What**?
Mind only.

ANYONE: *Mind only? You keep saying that, but what is that supposed to mean?*
All phenomena—all experience—is mental.

ANYONE: *"Are" mental, you mean.*
What we call physical phenomena is of a singular nature and mentally derived.

ANYONE: *What do you mean, "of a singular nature"?*
No separation of experiencer from experience.

ANYONE: *You mean . . . ?*
All is *empty* of intrinsic substance, of particularness.[94]

ANYONE: *So, when you say it's all just mind . . . what? You mean this is all like a computer simulation?*
Notice your knee-jerk impulse to put an object there.

ANYONE: *Put an object where?*
Like the idea of a computer, or some other thing, that's generating This. You never entertain the possibility that it's only *just* This from the get-go. Now you're going to have to account for all the computer hardware, its location, and speculate on who made it, and so on.

ANYONE: *So then . . . what? You're saying that this is all the mind of God, or something?*
Notice that you're still doing it.

ANYONE: *Doing what?*
Reaching for Something. You've injected the idea of God in place of a master computer. You don't need to do that. It only perpetuates confusion.

You don't need to create yet another mind object, including yourself. *Just see* thoroughgoing flux as Reality.

PART II

GRAND SYMMETRY AND
GRAND DELUSION

Religion short-circuits the religious experience by
putting it into concepts.
—*Joseph Campbell*

Let go of the idea "I exist."
—*Ikkyu*

27. WHAT ABOUT GOD?

ANYONE: *So, you're saying that even God is insubstantial?* Why bring in notions of God?

ANYONE: *Well, this highfalutin'* **mind** *you keep bringing up has got to be the mind of something! Why not call* **it** *the mind of God? You've got to have* **something** *there for it to be the mind of.*

Why? That's just another formed entity—another concept.

Why, especially, when every time we cling to concepts, we're ultimately left with contradictions? Worse, as in this case, we're left with something that makes us say with a straight face that it's all a big mystery, even when in fact it's all quite clear?

Why hang on to such a notion when it leaves your mind riddled with vacuous images, dubious arguments, contentious standpoints, and intractable theodicies?

ANYONE: *With what?*

Theodicies have to do with the problem of evil. They are futile attempts to vindicate God's providence, goodness, and power in light of the appearance of evil in this world. Given a substantialist's view, though, no such defense can be mounted without creating a raft of internal contradictions.[95]

Confused by thoughts, we experience duality in life. Unencumbered by ideas, the Enlightened see Reality.
—Huang-Po

ANYONE: *Why do you say that?*

You can't get these three qualities—all-good, all-knowing, and all-powerful—into a single entity. In fact, just the quality of being "all good" is, all by itself, incapable of fitting into a single entity. In any case, we've already *seen* that "entityhood" is illusory.

ANYONE: *So what's your problem with "all good"?*

Who can say what's good or bad?[96]

ANYONE: *Oh, come on!*

Just look at the run-up to the 2016 U. S. presidential election. One candidate was given hundreds and hundreds of hours of free air-time that none of the other candidates received. When asked about this, the head of one of the

TV networks admitted that, though this wasn't good for the nation, it was good for his company.

But **was** it? Was it good for ANYONE? Including the network executive—or, for that matter, even the candidate? Was it good for the world?[97]

The people who flew planes into buildings in New York and Washington, DC, and into the ground in Pennsylvania on 9/11, 2001, likely believed that they were doing good—and that they were serving God, as well.

So, what is good? What is evil? And who would you want to make that determination?

And which of these "goods" does God exemplify? Ostensibly, all of them, including the contradictory ones, such as killing people in the World Trade Center on 9/11 **and** saving people from those efforts to kill them.

Just today, I heard a news item about people caught up in one of the current conflicts in the Middle East. They were told that it was good and right to slay every "enemy" man, woman, and child; to slaughter all of their enemy's livestock and pets; to burn all their crops; and to add salt to their fields.

These concepts of goodness are identical to commandments that were ostensibly regularly issued by God to His people millennia ago—at least according to the Holy Books of the Middle East.

So, was all this good when God commanded it many centuries ago? Is it good now? Or is it evil? Or both? Or neither?

And how long do you suppose we could discuss just this single issue before breaking into conflict—or mutual tears?

Start making a list. You don't need to limit it to world events. Even in your personal life, I'm sure you've seen "good things" seemingly turn into "bad," and "bad things" "turn into" "good." I can think of events in my own life that seemed tragic at the time—yet seemingly favorable, life-changing results appear to have flowed from them. I've seen the reverse, too. No doubt you have as well. And these turnings never end.

As long as we believe that the things we think and hear and see and taste and touch are Real—i.e., substantial—we remain caught in one conceptual duality after another. And one conflict after another.

As for constructing theodicies in attempts to reassure and comfort ourselves, they always fail. In fact, they only deepen and extend our uncertainty and confusion.

Such problems, however, would never arise if such ill-founded assumptions were never made in the first place.

ANYONE: *Hmm. You said "assumptions." What other assumptions are you talking about?*

Take the common notion that an all-good, all-powerful, all-knowing God could also possess will.

ANYONE: *You're saying that God can't have will?*
Yes.

> I distrust those people who know so well what God
> wants them to do because it does, I have noticed,
> conform to their own desires.
> —*Susan B. Anthony*

ANYONE: *Why in heaven's name not?*
Would you say that God is Absolute and unlimited?

ANYONE: *Yes. Of course.*
The Absolute does not—and cannot—have preferences or will. Those are characteristics of limited beings, like us. We just project them onto our notions of God.

ANYONE: *How does that follow?*
There could be nothing outside "It"—an "Absolute Being"—for "It" to direct "Its" supposed will toward or against. In fact, such an ostensible "Being" would not have any attributes at all, including preferences or will, because those would relate to what is external to It.

ANYONE: *Well, if it can't direct its will toward things outside itself, what about things inside itself, then?*

We're talking about Totality. Absolute. There is neither "inside" nor "outside." If there were, your "it" wouldn't be Absolute. It would just be another concept. There simply **is** no "inside" or "outside" to Reality, or to Mind.

ANYONE: *Okay! So, then there* **is** *no difference between God and this high and mighty mind you keep mentioning.*
Mind, Totality, Wholeness is not an entity. There's no edge, no boundary, no division, no dimension, no location. We're attempting to speak of sizelessness, timelessness, seamlessness.

ANYONE: *Why can't we call that God?*
It's not just the notion of "God" that's being singled out here. **Any** ostensible entity is a concept, an illusion. Nothing more than mere appearance, mere thought construction.

Just stop assuming substantiality.

At the same time, however, don't ignore that This is not Nothing.

ANYONE: *Okay. So . . . ?*
Clinging to **any** entity—including self, other, cat, or God—puts us into a conceptual prison. This is true of **all** beliefs, all grasped concepts. They all **obstruct** True Religious Experience—direct experience of Reality.

ANYONE: *Wait. Isn't religion based on belief?*
As we'll see, religion is undermined by belief.

In any case, if you equate religion with belief, then you have to qùalify atheism as a religion, too.

ANYONE: *How do you figure that?*
It's belief-based.

ANYONE: *But atheists are non-believers!*
I've never met an atheist who didn't believe there is no God.

ANYONE: *Right.*
So, atheists are believers as much any theist. It's not that one is a believer and the other is not. They both believe. They just believe opposing points of view.

ANYONE: *So, what do* **you** *believe?*
I've told you, I don't hold any beliefs.

ANYONE: *Kinda wishy-washy, then, eh?*
No. I don't hang onto doubt, either—which is the flip-side of belief.

ANYONE: *Oh, because you've got it all worked out, I suppose?*
There's nothing to work out.

ANYONE: *So then . . . what? We just sit alone in the dark and endure the eternal barrage that emanates from Charles Ives's infernal trumpet? Is that what you're saying?*
Not at all. We all *know* perfectly well what's going on.

ANYONE: *We do?*
Yes. It's just that most of us ignore Reality. In fact, we go out of our way in our attempts to turn from It. We'd much prefer to embrace our cherished beliefs instead.

And that is our biggest problem.

28. BELIEF IS THE CULPRIT

Virtually all human conflict could be eliminated in a flash if we would simply abandon belief.

ANYONE: *Good luck with that.*
True Religious Experience demands it.

ANYONE: *How can you say that?*
Belief gets in the way of direct experience—religious and otherwise.

ANYONE: *But you can't have religion without faith.*
Who said anything about faith?

ANYONE: *Isn't that what you're talking about? Christian faith? Muslim faith? Jewish faith?*
You're confusing faith with belief.

ANYONE: *Faith* **is** *belief!*

Though the terms are often used in this way—or as two names for the same thing—they're actually very different.

Belief is a held view, and it always involves concepts and thought. It's what we think, or hope, or imagine, or guess is True (with a capital T).

Faith, on the other hand, simply refers to a sense of trust or confidence. And while we can have faith—trust, confidence—in what we happen to believe, faith is not confined to the realm of concepts.

Every belief—that is, every view held as True—is delusion simply because it's conceptual. Thus, as we've noted, any concept, any formed mind object—a unicorn, say—is illusory. We've also seen that **believing** that the formed mind object is Real is delusion.

Most people don't have trouble with this observation as long as the example used is a mythical beast. But we can go further.

We're talking about **any** formed object—a cat, say. It's illusory. Believing the formed object is Real—in other words, granting substance to that manifestation of Mind—is delusion.[98]

We can go further.

Any formed object—God, say—is illusory. Believing the formed object is Real is delusion.

ANYONE: *So, you're saying there's no God?*

Can't say there is no God; can't say there is. The same is True regarding a cat—or "you," for that matter.

ANYONE: *But if I believe in God . . .*

Then you're stuck in the Grand Delusion—believing that a concept, an object formed in the mind, however exalted and widely accepted, is Real and substantial.

ANYONE: *So, all belief is a form of delusion?*

Not a **form** of delusion. This is what delusion **is**.

And our deepest and most enduring delusion is our resolute belief in substantiality.

ANYONE: *So how do I get past this grand delusion?*

Suspend belief. Let go of it.

ANYONE: *Suspend **belief**? I've heard of suspending **disbelief**, like when you go to the movies. But how can we live without belief? How can you even walk down the sidewalk without belief? You said yourself that we all have unexamined assumptions about whether the sidewalk will hold us up, and so on.*

Just walk.

ANYONE: *Yes, but you have to believe that the sidewalk will hold you.*

Actually, you're probably not thinking about it, and you don't need to.

ANYONE: *But you trust that it will.*

That's faith. But notice: unlike with belief, when it comes to faith, thought doesn't need to be involved.

29. TRUTH OUTSIDE OF WORDS

Ultimately, we can't say anything regarding Truth or Reality. Truth simply won't go into concepts or words.

Nevertheless, we *see* Truth, directly. We experience Truth. In fact, Truth is **all** we ever experience. But most of the time we're lost in our thinking and not *paying attention* to actual experience.

ANYONE: *So, if truth isn't conceptual, are you saying we shouldn't even discuss it?*
Though we can't capture Truth in words, we can call *attention* to It, allude to It. We can use words to point.

ANYONE: *How?*
Consider fiction. Notice how it often draws our *attention* to Truth in ways that factual accounts cannot.

ANYONE: *Give me an example.*
Look at all the ancient myths about The Flood.

ANYONE: *I wouldn't have thought you to be a believer in* **those**!

I'm not.

ANYONE: *So, what's your point?*

Stories of cataclysmic floods have been told in myths and folklore around the eastern Mediterranean for at least 7,000 years.

ANYONE: *Then you* **do** *believe they're real!*

Stay with me, please. Geologic records provide evidence that floods began occurring all over the globe due to the melting of the great continental glaciers at the end of the last Ice Age, around 15,000 years ago. The subsequent period of sea level rise continued until around 7,000 years ago, when the Earth's global climate went into a period of quasi-stability.

There is a great deal of evidence from geology, archeology, and even linguistics that, during that episode of warming, a massive and devastating flood occurred around 5550 BCE, resulting in the creation of the Black Sea.

That event no doubt inspired many, if not all of The Great Flood stories that have come down to us since— fantastic tales of Gilgamesh and Utnapishtim, of Deucalion and Pyrrha, of Ziusudra, of Atrahasis, and, of course, of Noah.

These are wondrous stories that reveal a great deal about us as humans and about a spectacular event that some of our forebears witnessed. But the stories reveal these things to us only if we don't take them literally—if we don't confuse them with actual Truth.

So, to respond to your comment earlier, I deeply appreciate these stories and acknowledge their value—but I don't believe they are True. To believe them literally, we'd have to become as innocent and naïve as the imaginative people who created them.

> **If religion were merely an explanation, it would not show us the full range of human experience.**
> **—Dainin Katagiri[99]**

ANYONE: *But what value do they have if we don't think they're true?*

As our ancestors lived through such a cataclysmic event, they struggled to make sense of what they witnessed. In the process, they passed along epic tales and legends filled with such incredibly rich and heroic stories that, even from so many millennia ago, we still revel and delight in them today.[100]

These stories give us rich insight into their minds—into how they thought and how they understood their world. These insights also give us a fuller understanding of ourselves and our world today.[101] What stories **cannot** provide, however, is True Religious Experience.

30. THE UNWHOLESOME
NATURE OF BELIEFS

Have you ever noticed how beliefs diminish experience?

ANYONE: *Uh, no. Can you give me an example?*
Which would you say is the fuller experience? Comprehending the peculiar gait of astronauts walking on the moon because you've seen it on TV, or actually walking on the moon yourself? Imagining the taste of lemonade because you've heard that it's tangy and has a zesty fragrance, or actually smelling and drinking it?

> **It is that which you see before you—begin to think about it and you at once fall into error.**
> **—Huang-Po**

Having an idea, an impression, or a belief isn't remotely equivalent to having immediate, direct Knowledge and experience.

ANYONE: *But we've got to have beliefs!*
See if you can notice, then, how clinging to beliefs inevitably sets you up for disillusionment.

ANYONE: **What**? *Belief* **keeps** *us from disillusionment!*
Actually, you can't be disillusioned **without** belief.

ANYONE: *How can that be?*
What would you call disillusionment, then, if not the dashing of beliefs?

ANYONE: *Yes, but that only happens when a belief turns out to be not true.*
Exactly. And no belief is ever ultimately True, given its conceptual nature. Yet Truth doesn't require belief.

ANYONE: *But we still need beliefs. What about moral and ethical beliefs? If people didn't believe that it's wrong to kill, lie, cheat, steal, and so forth, we'd all be running around lying and cheating and stealing.*
Consider all the lying, cheating, stealing, and killing **already** taking place among us. Today, we in the United States routinely suffer more than 15,000 gun deaths per year—even though most of us believe that most of those killings were immoral. As for lying, cheating, and stealing, pick up any newspaper. Almost every day, some of the stories will be about leaders and other powerful peo-

ple who were found guilty of some or all of these unwhole-some practices. Belief clearly didn't stop them.

> **Believing what we don't believe does not exhilarate.**
> —*Emily Dickinson*

Belief in no way guarantees honesty and decency among people. In and of itself, belief is powerless to bring us to a just and peaceful world. Instead, our beliefs regularly lead us into conflicts that stir up humanity's worst injustices, sorrows, and pains.

As for the common notion that belief in God is necessary in order for us to do good, consider all those people who openly do not believe in God, yet basically live moral lives—in many cases, **highly** moral lives. And, of course, there are always some devout, godly people who conduct themselves in abominable, highly unwholesome ways.

ANYONE: *But don't we have to believe in* **something** *in order to be morally upright?*
No. Acting wisely and morally comes out of *seeing*, not out of believing.

Because belief mires us in the conceptual, the fractured, the divided, it also encourages us toward the small, the petty, and the contentious. The foundation of morality is not to be found there.

ANYONE: *So, on what basis can we find a foundation for morality if we don't rely on what we believe to be good or bad, right or wrong?*

Morality comes out of *seeing* Wholeness—which doesn't require any underlying belief. Indeed, belief puts up barriers to *just seeing*.[102]

Good and bad, right and wrong are concepts. Wholesomeness—Awareness of the Whole—is not.

ANYONE: *Wait. You used the terms "wholesome" and "unwholesome." Aren't they a pair of opposites? Aren't they relative?*

The Whole, of Its very nature, includes everything. It has no opposite. It no more dukes it out with unwholesomeness than the sky fights with clouds.[103]

31. TRUTH DOESN'T BELONG
TO ANYONE

ANYONE: *So, what are you, exactly? An atheist? An agnostic?*
Not atheist. Not agnostic.

ANYONE: *Really?*
To not hold views doesn't mean to not *know*.

ANYONE: *Oh, I get it. You're a mystic!*
No.[104]

ANYONE: *Well, what are you, then?*
Must you have a label?

ANYONE: *I just Googled you. I see that you're a Buddhist priest. So,* **that's** *your game. You're trying to get people to believe in Buddhism!*
Not at all. I don't believe in Buddhism.

In fact, I don't know what you mean. We've already seen the limitations and dangers of belief—and the damage it can cause. I'm not trying to get anyone to believe in anything. On the contrary, I'm hoping we can all loosen our grip on what we believe.

I recommend that you go easy on the word "priest" as well—at least when referring to what you think **I** am. I have no idea what I am. And I don't find it necessary to. I'm not anything in particular, and neither are you.

I'm simply talking about *waking* to Truth here. I'm not writing about Buddhism—or any particular religion, culture, identity, or angle. Reality doesn't fit into any concept—which includes any religious identity.

Truth has no religious affiliation. Gravity is not an Anglican idea just because Newton, who was born into an Anglican family, first described it.

ANYONE: *But Buddhism* **is** *a belief system.*
Maybe so. But what you're calling "Buddhism" isn't what the Awakened teach. It's not even what the historical Buddha taught. All that stuff came along later. And it's not a system that I believe in or formally follow. I don't hold any beliefs, "Buddhist" or otherwise.

ANYONE: *The awakened? Who are they, exactly?*
Buddha means Awakened—not asleep to Reality, not confused about what's going on. And Buddhism has no particular claim to such people.

As for the historical Buddha—a man called Gotama who lived in India 2,500 years ago—he was concerned with the nature of suffering and its release. He wanted to help people *wake up*. But he was just a man, not a god or a savior.

ANYONE: *But how do I know this book isn't an attempt to get us to follow the Buddha?*
I don't follow the Buddha, or anyone else—so why would I ask you to?

In fact, the Buddha never asked anyone to follow him, either. He said, "Be a light unto yourself." In any case, attempting to recruit people doesn't suit *Buddhadharma* at all.

ANYONE: *What's that?*
Buddhadharma is what Awakened people teach, regardless of tradition or background. The word literally means "the teaching of the awakened."

ANYONE: *But don't you believe that the Buddha was a prince, born to wealth and power, which he renounced in order to find enlightenment, which he did after working on it for countless lifetimes . . . ?*
I believe no such things. As I said, I don't hold **any** beliefs.

ANYONE: *You believe in reincarnation, don't you?*
Nope.

ANYONE: *But you have to if you're a Buddhist!*
I never said I was a Buddhist. You did.

Some religious scholars and historians argue that the man we call the Buddha went out of his way to free people from a belief in reincarnation—as well as from belief in general. He helped people *see* that there's no persisting thing—no particular self that endures from moment to moment, let alone from "lifetime" to "lifetime."[105]

ANYONE: *Then why do so many Buddhists believe in reincarnation?*
Good question. Honestly, I don't know. Maybe because that particular belief saturated the culture at the time of Gotama. Most of the evidence shows that he actually spoke **against** the notion of reincarnation. Instead, according to some religious scholars and historians, he spoke of what he called *rebirth consciousness*—of the nonstop reappearance of forms as though Reality were constantly being reborn, moment by moment, in seemingly continuous change—and of the equally continuous reappearance of mind and body in what we naïvely imagine as self—or, more intimately, as "me."

Gotama was speaking of conscious awareness, not reincarnation. He was noting that *this moment* looks very much like—but not identical to—what immediately preceded it.[106] He was examining Reality and consciousness and Awareness—just as we've been doing in this book. He was also examining ever-present change, and time and motion, and cognition and conception—also just as we have been doing.

ANYONE: *But you have all those stories and scriptures in which the Buddha talks about his past lives!*

Another good point. These were likely written down much later by others who came along generations afterward and attributed them to Gotama. Such tales and fables had long been in the culture even before Gotama's time. There's no compelling evidence that Gotama himself spun such tales. Certainly, they don't comport with the subtle, practical, and down-to-earth teachings we otherwise attribute to him.

ANYONE: *Like what?*

Like when he said, "*See* and *know* for yourself. Don't take my word for it. Don't take things on hearsay, either, or because you've read them in a book, or because other people believe them."

ANYONE: *Then why should I accept what **you're** saying?*

I'm not asking you to accept anything. Just *see* for yourself.

ANYONE: *Hmm. I'm still unclear about the distinction between rebirth and reincarnation.*

We don't know precisely what Gotama said, since nothing was written down until well after his death. But though "rebirth" is commonly taken to mean "reincarnation" these days, the fact is that this interpretation flies in the face of Gotama's central insight—that we don't actually experience a persistent self.[107]

Any notion of reincarnation begs the question of what is supposed to be reincarnating. And, though the question has been hotly debated over centuries, no one has ever offered any coherent justification for how the notion of reincarnation fits with what appears to be the main thrust of Gotama's message.

ANYONE: *But you **are** a Buddhist, right?*

I have never made that claim.

Buddhish, possibly.

ANYONE: *But you teach, and you've published books, and you've been ordained as a Zen Buddhist priest.*

Yeah, and received Dharma transmission.

ANYONE: *What's that?*

An endorsement to teach.

ANYONE: *By whom?*

By someone else who was given permission to teach before me. I wouldn't put too much emphasis on it, though.

ANYONE: *Why not?*

You need to *see* for yourself.

I'm only attempting to help you *wake up*. I don't care whether you embrace Buddhism, or reject it, or do some of both.

ANYONE: *I still think you're a Buddhist.*

If you must. But the historical Buddha never claimed to be a Buddhist.

ANYONE: *Well, of course not!*

He also didn't claim to be any particular sect.[108] Nor did he say that we should believe in him—whatever that would mean. In fact, he said something quite different: "*See* for yourself."

ANYONE: *Yeah, see and believe.*

No. *See* and *know*. *See* **instead of** believe. As we've seen, belief is a hindrance to Awakening.[109]

ANYONE: *You practice meditation, don't you?*
Yes. For more than fifty years, now—daily.

ANYONE: *And you write.*
And teach, to help clarify what the Awakened have pointed out.

ANYONE: *No study?*
Of course, daily, but that's secondary.

ANYONE: *What's first?*
Practice.

ANYONE: *You mean meditation?*
Yes. If we aren't talking to ourselves, if we simply *pay attention*, we can *see* what's going on.

ANYONE: *Hmm.*
That's the proper place for faith.

ANYONE: *Faith in what?*
Reality.

32. RELIGION WITHOUT BELIEF

ANYONE: *So, there* **is** *a place for faith?*
Of course. Just don't confuse it with belief.

ANYONE: *What's your take on an atheist like Sam Harris, then? He spurns faith in his book* The End of Faith. *I think he identifies with Buddhism, too. At least, sort of.*
In that book, he profoundly confuses faith with belief. Of course, he's not alone. Most people do.

While the book does a fine job of outlining many of the problems that religious beliefs entail, his remedy for fixing them resorts to the same tactics that he reviles in others.

ANYONE: *Namely . . . ?*
He falls back on belief.

His solution for getting humanity out of the bleak and senseless situations we repeatedly put ourselves in actually preserves the very problem he wants to end. He's as much caught up in belief as those he criticizes. Like most people, he's let his beliefs override his ability to descry.

ANYONE: *Descry what?*
Reality.
Just This.
Peace and understanding lie with Knowledge—never with belief.

ANYONE: *So, you're saying that because religion handles belief badly . . .*
Religion can't handle belief **at all**. Look at all the horrors and barbarities religion weaves from belief.
The saddest part is that religion doesn't **need** belief. It simply needs to open us to Reality.

ANYONE: *So, okay, religion can't handle belief. Is there anything that can? I presume your answer is no.*
Actually, science can, if it does so gingerly.

ANYONE: *But science isn't about "belief"!*
It certainly is. Rational, well-formed, scrupulously tested beliefs. That's essentially what scientific theories amount to. Beliefs formed after following strict rules and procedures that involve systematic and repeatable observation and measurement—and the formulation, testing, and modification of hypotheses. All subject to peer review.

So, unlike religion, science provides us with a valuable way to relinquish beliefs when they turn pernicious, or turn out to be completely out of step with Reality.

ANYONE: *But you told me to* **stop** *hypothesizing!*
If you want to *wake up* to Reality, yes. That's what you need to do.

If, on the other hand, you're interested in creating practical, useful conceptual constructs and tools that may (or may not) help you navigate everyday life, then look to science, and to the scientific method.

ANYONE: *Can't we do both?*
Yes—but only if you're mindful of what you're doing and you don't confuse the two.

In fact, this is precisely what we need to do.

33. JUST NOTICE AND RETURN

ANYONE: *So, you're saying that beliefs don't always get in the way?*

Beliefs always muddle our otherwise direct experience of Truth. But if you hold them tentatively and lightly, test them out, and relinquish them when they fall short—which they inevitably will—then they can sometimes be useful in limited ways.

But we need to be careful and discerning when it comes to forming and using beliefs. We need to notice when we're gripping them too tightly.

ANYONE: *And how do I know when I'm doing that?*

It's a matter of *attention*. Just watch your mind.

ANYONE: *Get my mind under control, you mean?*

Strictly speaking, it's not actually **your** mind, remember. And it's a matter of Awareness, not of control.

ANYONE: *Awareness of what?*
Nothing in particular. Just the Whole.

ANYONE: **Just** *the* **Whole?** *You mean* **everything?** *How can I be aware of all that?*
It's very simple, really. Just notice whenever your mind is inclined toward "this" or away from "that." Eventually you'll realize that this mundane awareness is equivalent to *seeing* the Whole.[110]

ANYONE: *Huh?*
Just notice when you're angry, or when you've begun to tighten down on a position, or when you're taking yourself too seriously, or when you've lost sight of the fact that you're only holding a mental model. Notice when, in your mind, you've frozen some aspect of the ever-changing World—including yourself—into a concept or an entity. Notice when you **think** you've actually gotten hold of Something.

ANYONE: *And then?*
Return to what appears *here, now.*

34. LEAVE BELIEF TO SCIENCE

ANYONE: *So, why are beliefs helpful in science but not in religion? I'm still not clear on this.*

Science has a fairly good record of reexamining, refining, and letting go of beliefs. This is a major way in which science differs from religion—at least religion as it's commonly practiced.

ANYONE: *Well, then, how can we trust scientific pronouncements?*

You mean put our faith in them?

ANYONE: *Sure.*

It's precisely because science **can** and occasionally **does** tear down its own edifices that we can put some faith in it—that is to say, a limited amount of trust and confidence. Because this is how legitimate science works, we can have faith that most unworkable ideas will eventually be weeded out.[III]

There will always be people chipping away at the margins of accepted knowledge. Some of their ideas will turn out to be helpful, some pernicious. And plenty of them will be happy to prove that an accepted hypothesis contains an error. So, even though revolutionary and penetrating insights and original ideas are often overlooked for decades, or even centuries, science has a means to eventually reconsider them.

ANYONE: *Or maybe not.*

Or maybe not. Science is far from perfect, and it suffers from all the typical problems of all institutions and systems created by people. Still, unlike institutions rooted in rigid belief systems, science never completely shuts the door on new ideas—or even old ones, for that matter. For science, the story is never over, and the door is always kept open—or, at least ajar—to new discoveries and challenges.

Time Out!

To illustrate this with one stupendous example, let's look at Alfred Wegener, a brilliant German interdisciplinary scientist. Like many others before him, he noted the remarkable fit between the west coasts of Africa and Europe and the east coasts of the Americas. After compiling copious amounts of data from the fossil record and geological formations on both sides of the Atlantic, Wegener proposed the notion of "continental drift" in his first edition of *The Origin of Continents and Oceans* in 1915.

His proposal was met with nearly universal ridicule from the science community. The problem was that, although Wegener had amassed a great deal of evidence to support his hypothesis, he could not account for any mechanism capable of moving continents around.

All of this was put to rest in the 1960s, however, with the advent of plate tectonic theory, which is now seen as one of the most important geological theories of all time.

Plate tectonics describes the structure of the Earth's crust as divided into large sections of rigid lithospheric plates that float on, and slowly move over, the mantle beneath. Most of the Earth's seismic activity occurs at the boundaries of these plates, where they collide with each other.

Given the immense implications of continental drift—not only for geology, but also for geophysics, oceanography, paleontology, and other disciplines as well—no one laughs at Alfred Wegener any longer. In fact, today he is held in high regard by many scientists.

Unfortunately, he did not live long enough to enjoy his vindication. He died while on a rescue expedition in Greenland in 1930, a few days after his fiftieth birthday.

No doubt valid insights by farsighted visionaries are being overlooked and scoffed at today. Nevertheless, science has a mechanism in place that works to eventually separate the gold from the dross, if imperfectly—and sometimes only after decades or centuries of resistance. (Recall Aristarchus's insight that the Earth orbits the sun, which was derided and ignored for seventeen centuries before being reexamined and then widely accepted.)

Religions that base themselves on belief have no corrective mechanism like this. Once the Word is given, it is nearly impossible to deviate from it, let alone overturn it—even in the face of clear evidence that it's false.

To the extent that certain doctrines **are** overturned, it's almost invariably due to the influence of outside social forces that cannot be repelled or denied.

This is precisely why religion should not make belief a central pillar—or in any way rely on it.

ANYONE: *But still, isn't belief the ultimate purpose or function of religion?*
You don't give up, do you? You ask this question because that's the only sort of religion you're familiar with. If you look more broadly, though, you'll find that religion has

evolved many different forms. Some forms don't rely on belief at all.

ANYONE: *What would you say is religion's ultimate purpose, then?*

It's not that different from science, really. Both spring from the same impetus—our desire to *know*, to silence Ives's untiring trumpet. Both are ways we humans have devised to get at Truth.

But, unlike the generally plodding and deliberate pace of science, religion has the capacity to settle the matter in a stroke.[112]

ANYONE: *What? You mean by just proclaiming truth?*

No. Of course not.

ANYONE: *How, then?*

By encouraging people to *wake up* to immediate experience and to *realize* Truth directly for themselves. In directly *seeing* Reality, the Big Question immediately settles. Everything clears up.

> Everything is clear to those for whom nonsubstantiality is clear. Nothing is clear to those for whom nonsubstantiality is not clear.[113]
>
> —*Nagarjuna*

ANYONE: *How can that be?*

There is no Mystery.[114] You already *know* Reality. You only need to stop talking to yourself, and learn to **wordlessly** *pay attention.*

Of course, religion can't do this for you. It can only assist—but again, only so long as it doesn't get caught up in belief.

ANYONE: *So, religion* **can** *save us, but usually does the opposite?*

To the extent that religion makes us reliant on belief, it forfeits its unique capacity to direct our *attention* to the Genuine, the Real, the True. Instead, it dooms us to endlessly fight over petty concerns—and, most of all, over whose beliefs are right.

Belief is the problem. Sans belief, religion can directly point to Reality.

Belief is the bane of religion. It walls us off from what religion truly has to offer: the end of fear, hatred, and division; the end of confusion; the end of not *knowing* what is going on.

35. WHAT SCIENCE CANNOT TOUCH

Science cannot touch nonsubstantiality. As useful as science is, it will never provide a way for us to *wake up* to Ultimate Reality.

ANYONE: *Why do you say that?*
Science remains forever within the conceptual.

ANYONE: *Forever?*
It wouldn't be science otherwise. This isn't a criticism. It's a necessary and unavoidable limitation.

ANYONE: *So, what's wrong with the conceptual, again?*
Did I say there is something wrong with the conceptual? The conceptual is functional; it's useful. It simply can't express or embody Truth, that's all.

Science is at home in the conceptual. It belongs there. This is where it soars. It can't freely function anywhere else.[115]

But remember: this is also territory where belief lives and flourishes. And any religion that is reliant on belief will be found here as well.

Unlike science, though, religion is not tethered to the conceptual. Thus, it's quite capable of directing our *attention* to Truth—but only so long as it abandons belief.

ANYONE: *You make belief sound downright dangerous.*

Belief **is** dangerous—in anyone's hands. In the hands of religion, it easily leads us to carry on Crusades, to form ISIS, to justify inhumane treatment of others, and the like. In the hands of science, it entices us to grasp at powers that can easily lead us into disaster.

ANYONE: *Like what?*

There are innumerable examples, but here are just a few:

We've overused antibiotics so consistently that we can no longer keep up with the super germs that have evolved resistance to them.

Then there's nuclear waste. What are we to do with it? These materials will remain exceptionally toxic for another hundred thousand years.

And then there's global warming. Consider the flooded coastal lands, the torrential downpours, the extended droughts all occurring now.

Then there are the novel plant diseases that have quickly swept around the world due to widespread monoculture farming. And the social and political breakdowns due to stresses caused by environmental damage and the loss of food production.

We now have the technology to manipulate any cell's gene sequence by cutting it at any desired location, and removing and/or adding genes. Gene sequences from any plant or animal species can now be combined directly. We can literally design our own creatures, including our own children. A few positive traits from a fruit fly or a raspberry? Sure. Why not?

This is all profoundly dangerous.[116]

The problem is that science has no built-in brake pedal. And, although science accommodates curiosity, ingenuity, flexibility, and even self-correction, it reserves very little room for Wisdom—Awareness of the Whole. Wisdom speaks to the human heart in ways that scientific analysis can never reach.

In wildness is the preservation of the world.
—Henry David Thoreau

36. THE TWO TRUTHS

There are two truths. I've been alluding to them in various ways for some time now. Both refer to aspects of Reality—that is, to aspects of Mind. Taken together, they reveal a pattern that physicists have come to call *symmetry*.[117]

First, there's the conceptual—the conceived, the individuated, the divided, the particulate, the quantized. This is the fractured and multifaceted world of conceptual experience—the world of form and movement, time and space. It's the world of constant, thoroughgoing change. It's the familiar world that appears, not **to** consciousness but **as** consciousness.

This is reality-with-a-small-*r*.

Then there's the perceptual—the inconceivable, the singular. This is the timeless, dimensionless, seamless, unchanging World of Totality—the World as Awareness, the World as Mind.

This is Reality-with-a-capital-*R*.

The first is the realm of relative truths, where there appears "here" as opposed to "there," "large" as opposed to "small," "love" as opposed to "hate," "good" as opposed to "bad," as well as time and space, motion and rest, and all other conceivable contrasts and dualities.

The second is Truth not relative to anything at all in particular. For lack of a better word, Absolute Truth—without reference to (or the presumption of) relative distinctions such as location, size, dimension, change, or any other attribute that would seemingly belong to "things."

The first aspect appears dynamic; the second, eternal.

All beliefs—including all religious beliefs—and virtually all of science, including scientific beliefs—necessarily remain within the realm of relative truth, relative reality. Consequently, all beliefs are transitory and subject to contradiction.

Science is, to some extent, designed to handle this actuality. Indeed, coping with it is integral to its proper functioning.

Religious belief, on the other hand, falls apart in this realm. It has no sound way to handle the inevitable contradictions that appear. So, instead, it puts on blinders, plugs its ears, invokes mystery, and uncritically presses on like a drunken fool, oblivious to the perils it needlessly subjects all of us to.

ANYONE: *But science invokes mystery, too.*

Some might use that word, perhaps, but no overarching mystery is necessarily presumed. Though I'll grant you that science is fully capable of bringing on powerful experiences every bit as moving as religious ones.

ANYONE: *Like what?*

Some of us are old enough to remember how, worldwide, we humans were collectively stunned by our first view, ever, of the Earth rising over the moon.[118]

On a less publicized occasion, imagine how it must have been for the first people to gaze upon the Hubble Deep Field.[119] It is one thing to stare at the night sky and marvel at all the stars, but quite another to have the Space Telescope stare at a wee patch of sky unflinchingly for days, where nothing appears, even to our conventional telescopes, and to find millions of galaxies in that tiny patch. And then to look in the opposite direction, and find essentially the same overwhelming view—and, finally, to realize that the universe appears like this in all directions.

Even so, whatever science discovers or reveals, it can never proclaim that it has touched Ultimate Truth. Science can only uncover relative truths.[120]

Religion, on the other hand, because it is not limited by the conceptual, can aspire to what is uniquely suited to its nature.

ANYONE: *And that is . . . ?*
Awakening.

ANYONE: *To?*
Reality.

ANYONE: *You said "aspire." What do you mean?*
True Religious Experience.

ANYONE: *You mean like what people report while on psyche-delics—being blissed out, or seeing colored light radiating from plants and animals, or having a sense of awe and wonder, or feeling that you're one with everything, or that this is the most significant experience you've ever had?*
No. This is not what I'm talking about at all.

ANYONE: *What are you talking about, then? Describe it.*
Words will never capture It. But I'm not talking about anything so noisy and rapturous as what you've described. Actually, I'm not talking about anything in particular at all.

ANYONE: *Then how am I supposed to have this true religious experience?*
Just practice Awareness.

ANYONE: *I do practice awareness.*

You overlooked the word "just."

Just practice Pure Awareness. Objectless Awareness. Subjectless Awareness. Awareness without thought.

ANYONE: *And how am I supposed to do* **that**? *I mean, do* **you** *practice subjectless awareness?*
Slow down. Take a breath.

Realize what you've just asked.

It's a flat-earth question.

ANYONE: *So . . .*
Slow down.

ANYONE: *Yeah, but how do I . . . .*
You keep thinking, jabbering, grasping, expecting that you need to do something, or get hold of something, or figure something out.

Drop all that. Just settle down and notice what's going on. Notice your busy mind—all the chatter.

ANYONE: *I am aware of it.*
Sit down. Straighten your back. Put your hands in your lap and cast your gaze downward.

ANYONE: *And now what?*

Just follow your breath. *Pay attention.* Be aware of breathing in as you breathe in; be aware of breathing out as you breathe out.

ANYONE*: So, I should . . .*
Stop!
    Just stop.

ANYONE*: But my mind keeps going!*
Notice that. Don't add to it.

ANYONE*: How do I stop it, though? Isn't that what you're telling me to do?*
I'm not telling you to do anything in particular.
    Just Awareness. No words.

ANYONE*: You mean . . . ?*
Stop.

37. SETTLING THE MATTER

Religion must not be about belief. When it is, it doesn't work.

Belief kills all prospects for True Religious Experience.

The point of the spiritual life is to realize Truth. But you will never understand the spiritual life or realize Truth if you measure it by your own yardstick.
—*Dainin Katagiri*

If the point of religion is for us to *awaken* to Truth—that is, for us to realize and "rebind" with Truth—belief will only bar the way, by turning everything into unanswerable questions and unsolvable mysteries.[121]

If you believe that God exists, you'll eventually claim God to be a mystery.

If you believe subatomic particles exist, you'll end up claiming them to be a mystery.

If you believe matter produces consciousness, you'll eventually speak not only of the mystery of consciousness, but of the mystery of matter as well.

If you live out of beliefs, or assumptions, or concepts, eventually everything becomes a mystery.

If you believe in Bigfoot, you'll end up declaring Bigfoot a mystery.

ANYONE: *But that's just a made-up thing! Who believes in that?*

You make my point.

ANYONE: *How?*

You don't believe in Bigfoot, so, for you, there's no mystery about it.

This is how it is with any mind object—with any individuated entity you believe to be Genuine, Real, and True.

ANYONE: *And so, you're saying . . . ?*

Just slow down. *Just look!* Don't grasp. Don't assume. Don't rely on belief.

Directly *see* nonsubstantiality in **all** appearances.

ANYONE: *This is starting to scare me.*

Notice that.

This fear is pretty common. As we humans draw near Awakening, we often become fearful and turn away. Or run away.

This occurrence is common enough that it is depicted by two, large, menacing, demon-like figures that stand on either side of the gates leading to some Zen temples. They are called Paradox and Confusion, the "guardians of Truth."[122] And they might seem to block our way to Truth. But they actually don't. They only frighten and hinder us when we cling to beliefs.

As we turn our *attention* to Truth, we inevitably discover that our ideas are incapable of capturing Reality. We're well along the Way if we get to this point, but it is also where many of us become confused and frightened, and back off without *waking* to Truth.

But if we're serious about realizing Truth, we need to not be deterred. We need to pass through Paradox and Confusion. We need to let go of what we believe.

In fact, apart from belief, there **is** no paradox. Because we cling to our concepts, there only **seems** to be. And in *seeing* this, there is no confusion, either.

ANYONE: *So, what am I to do?*
Just notice how this works. Realize that it's not necessary to hold beliefs at all.

Recognize that we've all had direct Knowledge of Truth and Reality from the first.

Just practice Awareness. Live out of that.

ANYONE: *And that leads to what you call true religious experience?*
No. You're conceptualizing again. You're thinking: "A leads to B."

Stop analyzing everything. *Realize* The Grand Illusion for what It is. Grand Symmetry.

ANYONE: *Grand symmetry?*
Not one; not two.

ANYONE: *You mean . . . ?*
Not is; not isn't. Not Something; not Nothing.

ANYONE: *But . . .*
The moment you would resort to words, or concepts, or beliefs, you kill True Religious Experience.

You need to *just see.*

ANYONE: *Then please, sum it all up for me one last time.*
The things and thoughts of this world appear compellingly Real. And for most of us, that's all there is to it. We believe in all of it. And so, believing, we live for a while, suffer because of our beliefs, and then die.

Some of us, however, are disturbed by such a bleak picture. So we fabricate and institutionalize all sorts of conceptual ways to respond to this apparently stark and dismal situation. These include various visions of heaven and hell, nihilism, narcissism, and quiet desperation.

Some people hurl themselves, head first, into belief. They may **say** that they "know," but they remain stuck in a vicious circle of concepts—while fears and doubts lurk nearby, or just over the horizon.

Others decide that no one can possibly know. Others go shopping, or to the bar, or immerse themselves in their careers, or surround themselves with other believers of similar minds, or otherwise distract themselves.

Meanwhile, they're constantly bombarded with all the mysteries and conundrums that spring from clinging to concepts. Most rarely question any of it. Very few question substantiality.

Still others substitute one form of apparent substantiality for another. They believe that Something—God, a "higher plane," a multiverse, "the astral realm," a supercomputer, "many-worlds," an alternate universe— is behind the façade of phenomenal experience. They imagine a bigger, better, more elaborate, or more ethereal form of substantiality beyond the nonsubstantial world we inhabit. As a result, they, too, remain stuck in concepts.

Others get carried away by the vision that "all is one." But this, too, is delusion.[123]

ANYONE: *But you say there's a way out of this situation.*

Yes. *Just look.* Just observe without assuming, without analyzing, without thinking, without judging, without grasping. Without saying anything to yourself.

Just begin to notice.

ANYONE: *Notice what, though?*

How the World appears as this and that, coming and going, here and there, you and me. Yet, at the same time, how the World **also** appears as undifferentiated, timeless, and unchanging.

Notice how physicality fades into Mind.

How all objects are manifestations of Mind.

How the Universe has no intrinsic size.

How beginnings and endings can't ultimately be found.

How we can't find boundaries between distinctions.

How space, time, and motion reveal themselves as Mind.

How consciousness actually functions.

How Mind is not an entity.

How all of our flat-earth questions evaporate if we *pay attention*.

How we don't actually have Something rather than Nothing.

How we have only This. Only Thus.

ANYONE: *Okay, so, finally, what is true religious experience?*
Realizing that these are not two.

ANYONE: *What are not two?*
The two truths.

ANYONE: *But that would be a contradiction, wouldn't it?*
That would be Ultimate Symmetry.

ANYONE: *So, you mean that, in the end, the absolute and the relative—what you call the two truths—are the same?*
Can't say they're the same.

ANYONE: *Then you're saying they're really different, after all?*
Can't say they're different, either.

Read this book again. With all that we've come through together, you might be ready for it now.

Of course, it will be a different book as you go through it once more. You'll understand it more deeply.

ANYONE: *Are you saying I don't have the smarts to have gotten it the first time?*
Oh, you're plenty smart. You've been an excellent inter-locutor. But *seeing* Reality doesn't require intelligence.

ANYONE: *What does it take?*

Patience, honesty, steady practice, and the determination to get to the bottom of things. Not that there's actually a bottom. Or things.

But you need to first set down everything you're carrying. That's the hard part.

Just pay attention.

It's getting late. I've gotta go.

ANYONE: *Wait! What should I say if someone asks me why there is something rather than nothing?*

Help them to see it as the ultimate flat-earth question.[124]

APPENDIX A:
THE TROUBLE WITH TRUTH
THEORIES

Since ancient times, philosophers have overlooked direct experience—feeling, instead, a need to wrestle, not with Truth, but with abstract **concepts** of truth. And they have very little to show for their efforts.

Over and over, they fruitlessly attempted—and continue to attempt—to construct valid truth theories, which they deem necessary in order to arrive at Knowledge.

Unfortunately, these theories inevitably get in the way of actual Knowledge. How could they not? We already *know* Truth directly. Looking to theories—to concepts, to abstractions—only distracts us from It.

While most philosophers correctly agree that you can't claim to have Knowledge unless what you're claiming to have Knowledge about is True, what they're **calling** Knowledge isn't actually Knowledge at all. It's belief. (Most attempt, without success, to shore up their "knowledge" as some qualified version of a "justified true belief"—a definition most philosophers regard as neces-sary, but not sufficient.) And this is the problem.[125]

The fact is that belief can never capture Truth, because belief resides entirely within the conceptual. And **this** is why all attempts to concoct valid truth theories have failed and will continue to fail.

Yet truth theories are utterly unnecessary, given that we already *know* Truth directly.

Thus, entangling ourselves in needless speculations and hypotheses only distracts us from what we already *know*.

When it comes to *waking* to Truth, Knowledge, and Reality, our chronic problem is not that we lack proper theories, beliefs, concepts, or definitions, but that we ignore actual experience. In order to avoid further quag-mires and FEQs, we simply need to *see* where we're chronically confused.

So, let's take a close look at one of these truth theories.

The *correspondence theory of truth* is embraced by more philosophers than any other, if only by a slight margin.[126]

Most of us, in fact, whether we're aware of it or not, operate out of correspondence theory's prime assump-tion: that truthful statements correspond to facts in the world. Claims are considered true if they correspond to observed facts—e.g., "A book is on my desk." My idea of "a book is on my desk" corresponds to the fact that, as I write this, there is indeed a book on my writing desk. I can see it. I can touch it. I can hear it when I flip through its pages. Were you here with me at this moment, you could confirm these observations for yourself.

This all sounds reasonable enough. So, what's the problem?

Consider what we've seen thus far. "Book" and "desk" are not substantial entities "out there." "Book" is a concept, not Reality. Likewise, "out there" is a concept, not Reality.

To realize Truth, we have to let go of concepts altogether. This includes not just truth theories and physical objects, but also ourselves. After all, "we" are as much a mental construct as anything else "you" might wish to point to in an effort to establish truth.

Yet pointing to entities of any sort always fails to establish Truth. Truth **can't** be established, nor does It need to be. It's already *here*. It only needs to be *seen*.

It's worth noting that a significant number of philosophers **reject** correspondence theory—but not for the reason I've just discussed. These philosophers argue that, unlike the example of pointing to a book on a table, there is nothing ANYONE can point to that would justify a statement such as "murder is wrong." And they're correct that no such thing can be found.

But they overlook the two truths. Ultimately, there is no such thing as thingness. We can't point to **anything**—

any concept—that would somehow justify **any** truth statement.

This doesn't mean, however, that we're at a loss to resolve moral issues such as how to regard murder. On the contrary, once we let go of truth theories, and our attempts to stuff Reality into concepts, and our efforts to shore up one concept with another, we can rely on direct perception alone. We can then *see* everything in terms of Wholeness, which completely frees us from any sort of moral confusion.

If we weren't ignoring the perceptual, we would never ask about what we could point to in order to justify a statement like "murder is wrong." We'd immediately *see* that casting moral questions in dualistic terms like "good and bad," or "right and wrong," throws us needlessly into conceptual quagmires.

In short, asking if murder is wrong is an FEQ. It's simply not a valid question, since "rightness" or "wrongness," is not the sort of thing that **can** be ultimately True.[127] Conceptual constructs necessarily leave out full consideration of the Whole.

Only in *seeing* any situation—that is, in relying on perception rather than concepts, and on Wholeness instead of conceptual distinctions—do we have a sound basis for morality.

It's precisely because *seeing* can't be packaged into a theory that it's lived and experienced, directly. Only in this way can we *see* a particular act as wholesome or unwholesome, and we can act in light of the Whole.

As long as we try to "figure out" morality rather than *just see*, we'll **never** find what we're looking for. All truth theories (like all other theories) presume substance—particularity, thingness, locality, existence, space, time. They don't recognize that these are simply appearances, mental constructs.

You only need to stop—and you do **need** to stop—and wordlessly *see* this lack of substantiality.

You don't need to form any theories or ideas about Truth. If and when you do, those ideas will separate you from It.

When such ideas arise, let them go.

Instead, just be still, and *know*.

APPENDIX B:
MIND AND CONSCIOUSNESS

Our concepts do not, and cannot, account for actual experience, let alone for Reality.

This realization is pivotal if we would understand the nature of subjective experience, conscious awareness, and, indeed, Mind.

To fully appreciate this, it's crucial that we slow down and seriously question our tacit assumptions about consciousness: that it's something we possess; that it's produced by our brains; that it provides some sort of read on an objective world "out there"; and so on. These unquestioned beliefs both drive and are driven by our current preoccupation with brains and computers—just as, in times past, humanity's view of the world was modeled after images of clocks and other gadgetry of the day. We need to let them go.

Part 1—The Hard Problem Lies
in What We Imagine

People have noticed since the middle of the nineteenth century that there appears to be a correlation between conscious experience and brain activity. Unfortunately, serious study of that correlation over many decades has brought us no closer to understanding mind or consciousness.

In fact, the more we investigate this apparent correlation, the more elusive any genuine understanding of consciousness has become.

Even so, undaunted, we double down and work harder at trying to capture mind in a jar, or analyze it in a laboratory.

Unfortunately, this is simply not the way to go about understanding the nature of Mind—i.e., Reality—and, hence, consciousness.

Through most of the twentieth century, most philosophers, scientists, and psychologists assumed that the human mind was a black box, impenetrable to any kind of probe or research. "We can only observe behavior, not mind," was the dominating view in the mid-twentieth century.

But then, in 1974, Thomas Nagel wrote a most influential paper titled "What Is It Like to Be a Bat?" (He was referring to the flying, furry animal, of course, not to the wooden implement used in baseball.) This notion seemed to capture the fancy of virtually everyone involved with mind or brain research. Suddenly, it seemed that the black box of the human mind had been shattered—and our modern, wide-eyed era of consciousness and brain research had auspiciously begun. At last, we could begin to peer inside the human mind—or so it seemed to many.

It was a heady time. I remember in 1988, while working on my first book, a friend of mine, a professor of physics at a major university, was shocked to learn that I was planning to point out in my book that Mind has primacy over matter. "What are you going to do," he laughed, "when they build a conscious machine?" With utter confidence, he assured me that "they" were on the verge of producing such a device. "Any day now."

That was the general feeling in 1988. And, sure enough, within a few years, Daniel C. Dennett wrapped it all up with his book *Consciousness Explained.* Mission accomplished.

Okay. So, what went wrong?

What is wrong is that, to this very day, just about everyone ignores that Nagel's title is a flat-earth question (FEQ). It's not **like** anything at all to be a bat—any more than it's **like** anything to be you. Consciousness is not an object to be explained or described. It's not a subject, either. It's Total.

Given our knee-jerk assumption of a self, however, we attempt to study mind and consciousness in a manner that resembles a naïve person who keeps opening the refrigerator door to see if the light is off.

Noting how easy it is for everyone to get sidetracked by irrelevant particulars, in 1994 David Chalmers drew our attention to the "hard problem": "how physical processes in the brain give rise to subjective experience."[128] (Easy problems, according to Chalmers, are those that fall to standard methods of cognitive science. They're "easy," essentially, because, they can be investigated in experiments that can actually be carried out—at least in principle. This, however, is not the case with the "hard problem.")

The "hard problem" isn't actually hard. It's impossible. It's just another FEQ based on groundless assumptions.

Physical processes in the brain have **never** been demonstrated to give rise to subjective experience. We only **assume** that they do—because there **seems** to be a correlation.

But, as everyone knows, correlation is not causation. And, in this case, most of us are oblivious to the fact that we've made that unwarranted assumption.

It's time to admit that our preoccupation with properly arranging matter in a brain—or in a machine—to **produce** consciousness, whether artificial or natural, is doomed to go the way of our searches for phlogiston, caloric, or the number of angels that can dance on the head of a pin.

It would be fun, of course, to continue speculating on the experiences of zombies and bats and what Mary, the colorblind scientist, actually knows; or to reflect on the musings of Minsky, or the tests of Turing, or the implications of Searle's Chinese Room.[129] But the fact is that, no matter how fast or encyclopedic you make Watson, the IBM supercomputer, Watson **has** no "inner life."[130]

Yes, consciousness does seem to be absent unless various parts of the brain are all "talking" to each other. And by blocking that communication we can produce a powerful (and, in certain contexts, useful) anesthetic effect. But this only demonstrates yet another correlation. It does nothing to explain consciousness (not that consciousness needs explanation), or to demonstrate how the brain supposedly produces it.

But all of this is moot. The fact is, we have no evidence that consciousness even **has** an origin, let alone that it originates in matter—though, as we've seen in this book, there's plenty of evidence to the contrary.

Consciousness will always remain a mystery for those who continue in their attempts to construct it from matter.

A radically different understanding of what is actually going on is needed.

PART 2—THE EASY PROBLEMS LIE IN WHAT WE CAN'T IMAGINE

It turns out that consciousness poses a thorny issue for quantum mechanics (QM).[131] Yet the kinds of concerns raised in part I of this appendix give virtually no attention to the issue.

So, instead of letting this thorniness hang out there, let's reel it in and take a good look at it.

When you believe in matter—that is, when you believe that the physical world is truly "out there," encapsulated in substantial little packets or particles known as *quanta*—much about that world no longer makes sense. Most senseless of all is the "mystery" of consciousness, which seems to be in the picture somehow.

The classic "double slit" experiment demonstrates the inseparability of consciousness and the physical world. Physicist Richard Feynman claimed that this experiment contains the "only mystery."[132] It runs like this:

If you shoot a single particle at a screen with two slits in it, it seems to go through **both slits at once, but without dividing**. And the only way you can keep a particle from acting in this way is to gain awareness of—that is, to look and see—which hole the particle went through. In that case, the particle will appear to have gone through either one slit or the other.

This experiment has been repeated many, many times and been described in all sorts of ways, but that's the gist of it. The bottom line is, it doesn't make sense—not if you believe in thingness, in substantiality, in quanta, or in particles. Thus, as no less than Albert Einstein put it, "Fifty years of pondering have not brought me any closer to answering the question, what are light quanta?"[133]

So who believes in such things? Most of us, probably. But certainly, particle physicists do.

There are other physicists, though, who subscribe to quantum field theory (QFT). Field physicists will tell you that there are no particles, only fields.[134] What particle physicists describe as a *point particle*, or a *quantum*—is described by field physicists as an impulse, a disturbance traversing a field. (There are multiple kinds of impulses, disturbances, and fields.)[135]

And so, naturally, when an impulse, fanning out across space, encounters a screen with two holes—since it's a disturbance spreading through a field rather than a point-like object moving along a particular trajectory— the disturbance simply passes through both holes. And so, from the field physicist's perspective, what's the problem?[136]

And what has any of this to do with consciousness?

A field physicist might tell you that it has **nothing** to do with consciousness. At least, that's what Rodney A. Brooks claims in his book *Fields of Color*.[137] For field physicists, quantum field theory not only dispels the mystery of the two holes, but is also "the most successful theory . . . in all of science."[138] It explains almost everything for quantum physicists.

But not quite.

As Dr. Brooks points out in his book, five mysteries or gaps remain to be filled in. The first of these gaps is "field collapse." The second is "consciousness."[139]

It's worth noting, however, that the first two "mysteries" that Brooks mentions are precisely the two that he ignores in his account of the experiment with the two holes. There's a good reason for this, of course— or, at least, for ignoring the first. As Dr. Brooks puts it, physicists don't understand field collapse, since quantum field theory has "no equation or mechanism to explain or describe the process." He then adds, "To tell you the truth, we don't really know when field collapse occurs." And, a bit later, "Field collapse is a mystery because QFT doesn't predict when or how a quantum will change its state or be absorbed."[140]

As it turns out, this first mystery or gap has everything to do with the second. In fact, they are virtually one and the same.[141]

Brooks' ostensible "mystery" isn't that a wave fanning out through a field can pass through two holes at once. It's not even that the collapsing impulse puts all its energy into a single point—which it appears to do in the very instant of the collapse, even if the disturbance is spread out across the galaxy, or half the known universe. (This is the part Einstein initially had trouble with, for obvious reasons.)

The "mystery" **seems** to be that, if someone happens to be looking, suddenly the wave interference pattern, caused by field disturbances passing through the two holes, will "collapse" into a double diffraction pattern. It is as if particles, not waves, were passing through either one hole or the other. (I won't belabor this point here. If you need more information, Google "double slit experiment" or consult my book *Why the World Doesn't Seem to Make Sense*, pp. 70–77.)

The **only** difference between, on the one hand, a wave interference pattern building up behind the screen with the two holes and, on the other, a double diffraction pattern building up instead, would **seem** to be that someone was looking. Someone saw—someone had awareness—that the "wave" appeared to collapse to a "particle" and passed through either one hole or the other (thereby creating the double diffraction pattern in place of a wave interference pattern).

The question would **then** seem to be—and it is often asked amid howls and objections—what is it about human consciousness that causes the wave, or the field, to collapse? Brooks doesn't know. None of his colleagues has any idea. They're baffled.

Of course they are—because this is an FEQ.

Field collapse has little to do with "human consciousness," whatever **that** is. But it does have everything to do with consciousness.[142]

Just because we don't have an equation or mechanism in quantum field theory that describes when or how field collapse occurs, this doesn't mean that it "just happens."[143] The experimental results are clear and obvious: what we refer to separately as "consciousness" and "field collapse" are not, in Reality, two different, distinct things. Indeed, they're not things at all—in much the same way that inertia and mass are not two different, discrete things. We can't have one without the other.

The "mysteries" of "consciousness" and "field collapse" clear up in the context of the Whole.

The issue is not whether or not "someone is looking."[144] Rather, it is that consciousness **is** the fabric of Reality. All is Mind.

To press beyond this point, it will be helpful to consider whether quantum field theory, as it is interpreted today, is really any more (or less) sensible than quantum mechanics. (The next few pages get a bit technical, so if you find the waters too deep or turbulent, feel free to skip to part 3 of this appendix.)

First of all, we need to understand that QFT is not about classical fields (e.g., gravity fields or electromagnetic fields)—the fields we're all familiar with and can picture in our minds. For example, when you see the apple fall from the tree, or when your hair stands on end as you bring an electrically charged comb or brush to your head, or when you see iron filings form a pattern as they're sprinkled on a sheet of paper placed over a bar magnet, all of these phenomena occur in classical fields—a gravity field, or an electric field, or a magnetic field, respectively. But classical fields and their effects on the things that feel their forces are **not** under discussion here.

We're talking about a quantum field. Unlike classical fields, you can't picture what's going on in a quantum field. No one can. A quantum field is inconceivable.[145] It is, however, perceivable. Indeed, such goings on is all that is ever perceived. And this gives us yet another clue regarding consciousness, as we're about to see.

In classical field theory, every point in space has a varying but measurable numerical value that corresponds to whatever force is acting upon the particles (the things) that feel that particular force. In other words, a force might be stronger here than over there. The Earth's pull of gravity is stronger near the Earth's surface than it is at some distance from the Earth. In an electric field, the strength of the field's force becomes greater around electrically charged objects, electrical wires, and so on. In short, in classical fields there appear to be objects moving through space and time, and at any point we can measure the forces that act upon these objects. In addition, those forces themselves can be visualized as they vary in space and with time.

All of this is easy to picture and conceive, because every item appearing in classical accounts is conceptualized into little, quantifiable packets, or conglomerations of said packets.

Not so with a quantum field. We **cannot** simply take what appears to go on in classical field theory and super-impose quantum field theory on it. Indeed, it's impossible to conceive of what is going on in a quantum field. We literally cannot picture it.[146] Consciousness does not function there—but objectless Awareness does.

Thus we have two aspects of Mind, or Reality: (1) the conceptual, or the function of Mind we call conscious-ness; and (2) the perceptual, or objectless Awareness. The first aspect manifests as classical fields, the second as the quantum field.

These two, taken together, reveal symmetry.[147]

Classical fields, on the one hand, and the quantum field, on the other, are two different accounts of One Inseparable Reality. In one account, everything is under the streetlamp; in the other, everything is in the dark. In one account, everything is potentially graspable and con-ceivable; in the other, all is perceived, yet inconceivable. In one, mountains are mountains and rivers are rivers; in the other, "mountains" are not mountains and "rivers" aren't rivers. These are the two truths: apparent fragmentation and undifferentiated Wholeness.

This symmetry, when *seen* directly, is the silencing of Ives's trumpet.

PART 3—SO, WHAT'S THE PROBLEM?

Let's review:

Part 1: Matter, particles, forms, quanta, material, brains, and the like, cannot account for Mind or consciousness. Yet consciousness accounts for them. Physical objects do not, and cannot, precede or produce Awareness. Awareness precedes and produces them—or, more accurately, manifests their appearance as conscious experience.

Part 2: Wholeness = objectless Awareness = Mind. (Or, more accurately, Wholeness ≡ objectless Awareness ≡ Mind. The identity symbol (≡) means more than mere equality. It means "is not other than.")

Thus, two truths.[148]

Awareness is All; consciousness Its fragmentation. Uniformity; unending change.

The inter-identity of these two reveals Ultimate Symmetry. This objectless World of undivided Wholeness is none other than the everyday world of this and that.

Thus, Mind; thus, Reality.

APPENDIX C:
THE FLOOD AND OTHER TRUE FICTIONS

We can't throw away or ignore relative truths just because they're illusory. They still show up. They still affect us—and we still affect and create them. That's why they, or at least some of them, are still called truth (with a small *t*).

Let me tell you a story that begins about 15,000 years ago. As for its veracity, it's real enough.

In huge, turning cycles, the slightly oscillating tilt of the Earth, the ever-so-slight change in the shape of its orbit, and the shifting occurrence of perihelion in Earth's orbit all brought the Earth out of a long-endured Ice Age.

The seas began to rise.

After the Ice Age broke, within a millennium, people in East Asia began to experiment with wild plant foods. Millet, in particular, became an important food source. In time, East Asians began to experiment with rice as well.

In Southwestern Asia, the climate became ever warmer and wetter as the Earth continued out of the Ice Age. Oak and pistachio forests began to extend across much of the region. Human populations grew as food became more abundant. The hunter-gatherers who lived in the Jordan and Euphrates river valleys found rich supplies of nuts and wild grasses, as well as game such as gazelle.

As food became more easily obtainable, extended communities began to form. Within these communities, social ranking became ever more starkly drawn, in the interest of maintaining order and providing for the fair distribution of food.

As human populations began to organize into social hierarchies, so too the spirit world appeared more to be ruled by greater and lesser spirits—gods, really. These became ranked and arranged according to various functions and factions, just as in human societies. (Though, of course, it was understood at the time that humans organized themselves according to how their proto-gods did it, not the other way around.)

This new way of seeing the natural order of things, though long in arising, developed quickly, and soon replaced the simple animistic take on the world that humans and their forebears had operated out of since primeval times.

Hunter-gatherers living in northern Europe at this time gradually moved their camps farther and farther northward. They relied on reindeer and other animals for food, clothing, shelter, and other essentials. By 13,000 years ago, some of these various folks began to settle along what would become the Baltic Sea.

Then, abruptly, about 12,800 years ago, the Earth's climate fell back into, essentially, a resumption of the Ice Age.[149] Within a few generations, glacial conditions affecting the entire globe returned, and human populations everywhere were profoundly altered.

This quick return to Ice Age conditions was brought on by the sudden dumping of vast quantities of cold freshwater into the North Atlantic. As the North American ice sheet had melted during the preceding period of global warming, there formed an immense lake, much larger than any freshwater lake in the world today.[150]

The lake first drained to the south, down a tributary to what would later be called the "Father of Waters," the Mississippi River. But as the ice retreated, it eventually uncorked a more efficient channel of escape through a gap in the Laurentian divide, along the north coast of the "Big Sea Water," *Gitche Gumee*—or, as we now call it, Lake Superior.

As the massive flow of freshwater from the giant lake poured into the much smaller Lake Superior, it cascaded down the chain of Great Lakes into the St. Laurence River (as it would later be known), and out to the north Atlantic. There it abruptly shut down the natural circulation of warm water being carried northward from the Gulf of Mexico.

In short order, Europe plunged back into arctic conditions. Within a few decades, the effects were also felt in Asia and Africa, and even as far away as New Zealand. In southwestern Asia, the return to Ice Age conditions brought on cycles of intense drought, causing drastic reductions in food supplies. This in turn put great pressure on the growing human populations of those regions to feed themselves. It was all to have a profound effect on the course of human events.

People had to find more effective ways to obtain and preserve food. They discovered that the grinding of grains into meal became an effective way of storing food. And when supplies became short, some groups began to experiment with deliberately tossing seed among the wild grasses they depended upon.

As they first practiced this method, they used the wild seed that they relied upon, only lightly scattering cultivated seed to supplement their harvests. But then they began to tend to their plantings, and thus began to alter their environment. It did not take long before it dawned upon some people that great potential lay in deliberately planting their own food.

Then, suddenly, after 1,200 years of Ice Age conditions, the Gulf Stream reestablished itself, and global warming resumed—now even faster than it had before.

Again, the seas continued to rise. By 11,000 years ago, Asia had separated from North America; the British Isles became isles and separated from Europe. The Baltic Sea, previously just a freshwater lake, became an arm of the Atlantic.

As the Mediterranean basin continued to fill, and its waters encroached upon populated areas close to the sea, occasional sudden floods occurred in various places along the coasts. And as the oceans gradually encroached upon the land, in Southeast Asia hundreds of thousands of square miles of low and level land slowly, over centuries, became submerged. Occasionally there would be sudden flooding as water poured into low-lying areas. In time, Borneo, Sumatra, and Java, as well as countless other elevated areas that are now islands, became isolated from the Asian landmass. Australia and New Guinea parted ways as well, as the rising sea gradually took the low and level lands between.

These changes were noticeable for generation after generation. During the normal lifespan of a man or woman, the sea would rise five feet or more. The people living there, of course, had no understanding of what was happening. They only knew that the sea was gradually covering lands they once inhabited.

Like people everywhere, they needed an explanation, a story to account for what was going on. And they began to tell stories of a great frog that swallowed up the mighty sea, only to spew it back upon the land. This account was orally passed down from person to person and generation to generation, even to our present day. It is a story told by people to remember that there were once vast lands belonging to their ancestors that were gradually taken by the sea.

To the west, the lands we now think of as the subcontinent of India and the island of Sri Lanka were joined by an exceedingly narrow strip of land. People could walk to the "island" from the mainland and back. The locals referred to it as the Bridge of Prince Rama, and it figured in many of the stories told by people who inhabited those ancient, but still connected, lands. In time, however, this wondrous bridge, too, slipped beneath the rising seas, to be forgotten except as myth and legend. But if you look on any modern map that records the ocean's depths, you can still see the Bridge of Prince Rama.

In time, as Earth continued to warm, birch forests encroached upon what had been treeless steppes in Europe. Eventually, mixed oak forests began to cover the land. The people there became forest hunters, while those farther to the north, along the coast of the Baltic Sea, became expert fowlers and fishermen.

As the vast continental glaciers continued to melt, water collected in three large low-lying depressions, two of which formed the Aral and Caspian Seas. Having no outlets, these would gradually become salty. The much deeper depression west of the Caspian, however, was to have a much different fate. The runoff collecting here fed a gigantic lake hundreds of miles long and at least a hundred miles wide—Lake Euxine. And it would have gone the way of the Caspian Sea had it not been for a singular, cataclysmic event.

By 7,600 years ago, farming and fishing communities had settled along the huge lake that lay just north of the Anatolian Plateau in Asia Minor. (Actually, "Asia Minor" was poorly defined as such, since there was as yet no Black Sea to give it its modern definition. There was just Euxine Lake, which covered no more than half the expanse of what would become the Black Sea.)

People were drawn to the lake, for along its shores there was good soil to support farming. The lake also supported a fishing industry, and villagers in the area made an industry of trade as well. They also kept animals, including pigs, since this was near where those animals were first domesticated.

All in all, the lake supported many thriving communities along its shores, perhaps as many as 150,000 people or more—a considerable number for the ancient world.

Sometime before the great event, a genetic mutation occurred among this population—a mutation that introduced blue eyes to the human world. Their language was Proto-Indo-European, the forerunner of present-day languages as diverse as Sanskrit, Greek, Russian, and

English. These people would eventually disperse their language, culture, technology, folklore, and genetic traits—and their gods—across much of the Western world and beyond.

The oceans of the world had been steadily rising for millennia, though so slowly that few people would have noticed the change, even over generations. Changes **were** noticed by some, though, particularly those in level areas close to the sea, such as the inhabitants of Southeast Asia and Australia.

Similar changes also crept into the Mediterranean basin. Sudden flooding occurred here and there as seawater gradually rose above local depressions or curled about to form islands, especially in the eastern shallows of that Great Sea in the middle of the Western world.

No doubt it was an exciting moment for any who witnessed it when seawater merged with the freshwater that once drained from the Gulf of Corinth—thus making that body of water into the Gulf of Corinth—or when seawater rose to join the Sea of Marmara, now no longer an outflow of freshwater into the expanding Aegean. But this was nothing compared to what was to befall the populations that lived along the shores of Euxine Lake, most of whom are now lost to memory.

As the oceans of the world, all joined as one body, rose, so too did the Mediterranean, itself but an arm of the Atlantic—and the Atlantic, along with the Indian, but an arm of the vast Pacific Ocean. All stood poised at the Bosporus, a plugged channel or fault traversing the narrow isthmus that held back all the oceans of the world, keeping them from spilling into Euxine Lake, more than 500 feet below the level of the Worldwide Sea.

Perhaps it was nothing more than two rivulets, one draining to the Sea of Marmara, the other toward Euxine Lake, suddenly connecting at their headwaters during a huge rainstorm. But more likely it was a jolt from one of the earthquakes that were (and still are) common in that region. Without warning, the plug was abruptly cut through to a grade below the level of the Great Sea.

Immediately the deluge sliced through the breach and down. Backed by all the oceans of the world, the mass of water surged into the depression, scouring out a channel two to three miles wide and hundreds of feet deep.

Though the oceans of the world barely registered their losses, their combined contributions produced the greatest natural disaster—at least in terms of sheer physical magnitude—that has ever been witnessed by humans. It was a surge of more than a thousand Niagaras, all suddenly pouring into an enclosed basin.

For the people living along the shores of Euxine Lake, there could be no understanding of what was taking place, or why. They knew nothing of melting glaciers or of Ice Ages; nothing of the Earth as a sphere rotating in space, revolving around the sun; nothing of the great oceans supplying the vast quantities of water that now surged into their world. They had no concept that they were living hundreds of feet below sea level, and that the seas were now pouring in. All they knew was that their world—their homes and villages, their fields and flocks, their fences and trails, their way of life, their families, friends, and neighbors, even their boats and canoes, were all being swept away. Their lands were suddenly, swiftly, inexplicably, and without warning, being covered by water, never to be seen again. To them it seemed the entire world was being taken beneath the waves.

People living hundreds of miles from where the flood was surging in would have no idea of why the lake suddenly began to rise. Nor could they have anticipated how high it would go. In a matter of days, it would rise hundreds of feet and stretch out across tens of thousands of square miles, never to recede.

The lake rose so rapidly that many people surely had difficulty escaping—especially if they tried to take anything with them. Many must have perished as they clung to patches of high ground, which in short order became submerged. Perhaps some took to their watercraft and survived; or others, out of sheer luck, managed to stay on the mainland as the waters rose.

But along a portion of the lake, about 400 miles to the east of where the deluge poured in, a ridge rose high behind the villages along the old lakeshore. Its summits towered hundreds of feet above the peoples' homes. Surely, people must have thought, as they scrambled up that range of peaks, they could stay ahead of the water and climb to safety.

But as the rising waters flowed in behind the ridge, the range itself was cut off from the mainland. Those who scurried up the ridge were now on an island (though they would not have known that at first). This newly formed island would have been at least fifty or sixty miles long when it separated from the mainland. But the waters just kept rising—and, as they did, the island became ever smaller and farther off shore.

In a matter of days, the waters neared the peaks of the former range. By this time, the island had broken into a series of smaller islands. Any survivors, still clinging to the peaks, would now be forty to fifty miles off shore as the waters finally overtopped the peaks and kept on rising.

It must have seemed unreal to those who were there when the waters submerged what they once knew as high peaks, leaving those, now swimming for their lives, at sea and out of sight of land. The former range had now become a seamount submerged off the southern coast of what had become, in only a matter of days, what today we call the Black Sea.[151]

Surely this event had an enormous impact on all who survived. For those who may have been rescued at sea, it must have seemed as if the whole world had gone under. For them, the world, though large, was not anywhere near as large as what we now know the World to be. For them, the sky, where the sky gods lived, was not that far away—just overhead.[152] And the gods, from their perch—could they not look down and see all that was going on? An explanatory cosmology was yet to be devised, but the minds of these people were becoming deeply etched with new and startling visions and ideas.

As for the earth, did it not rest upon abyssal waters? So some indeed came to think. After all, for those who lived through the cataclysm, the waters seemed to come from below—and they were great enough to submerge the world.[153]

These events—these unbelievable events—were indelibly impressed upon all who witnessed them and survived. Evolving variations and interpretations of The Flood would reverberate through folktales long, long after the event itself passed from living memory, for well more than a hundred generations.

There are not just the two accounts of Noah that appear in the Old Testament. (In one version, Noah brings onboard the ark two of every creature, while in the other he takes on two of each unclean creature and seven of each clean one.) There are many, many other tales as well. From the epic of Gilgamesh, we have the tale of Utnapishtim, who built a boat as wide as its length, with one full acre as her floor space, ten dozen cubits the height of each wall, and ten dozen cubits each edge of her six square decks. Utnapishtim brought onboard the seed of all living things on earth and floated about until coming to rest on Mount Nisir. Thereupon Utnapishtim sent forth a dove to search for land, but the dove returned. Next, he sent forth a swallow, who came back, also not finding any place to rest. And then he sent forth a raven, who saw that the waters had receded, and Utnapishtim offered a sacrifice and poured out a libation on the top of the mountain. And he and his wife were made as gods by Enlil, who blessed them with immortality, saying that Utnapishtim would reside far away, at the mouth of the rivers, and from thence was to be called Utnapishtim the Faraway. It was from Utnapishtim that Gilgamesh sought the secret of immortality, but instead learned the responsibilities of a king.

And then there were Deucalion and Pyrrha, who survived the floods that had been sent by Zeus to punish human wickedness. But Deucalion and Pyrrha were spared and instructed to throw stones over their shoulders, which turned into humans to repopulate the world.

The Greeks had other tales inspired by those cataclysmic times as well—such as that of Jason, the sailor, at the Bosporus. There was no way through to the Black Sea without passing between the clashing rocks that guarded the Straits of the Bosporus. These would close together and smash his ship as it attempted to sail through. But Jason learned from a blind prophet how to fool the rocks. He needed to send forth a bird ahead of him. The clashing rocks would then close on it, and when they reopened, Jason's ship would be allowed to pass through.

It's easy to see how such tales are based on older stories that came from shortly after the time of The Flood, when people saw how the rocks parted and opened the channel. Quite likely the rocks that remained standing at the Bosporus had been loosened enough to be unstable for years afterward. Perhaps, for a period of time, they continued to be tossed about by high seas and tides during storms—or rattled by later earthquakes or aftershocks.

Notice how these various details can be woven together into a single narrative. These details come not just from cosmology, geology, linguistics, and archaeological artifacts, but also from evidence richly scattered among stories told by people who were trying to make sense of what was taking place around them.

But, unlike the isolated, phantasmagoric tales told by the ancients, who had comparatively limited worldly knowledge, we can fairly easily put together a story that takes into account **all** the currently known facts—i.e., all the relevant relative truths. These include multiple far-ranging tales, such as those from a much earlier time, by people on the other side of the planet, who imagined a giant frog that spewed out water upon the land, and from people in South Asia who told of a bridge, long ago, that connected India with Sri Lanka. It all fits together into a single narrative.

Such accounts have continued down to the present day. As glaciers around the globe continued to melt, at one time huge torrents streamed from the Himalayas, and a great civilization grew and flourished along the rivers that drained that prodigious melt from the high peaks. We knew nothing of these people until roughly a hundred years ago.[154]

And then there is the great diaspora of the Proto-Indo-European-speaking peoples, which spread in all directions from lands that are no more—lands taken by the sudden formation of the Black Sea and later forgotten. It was from these scattered people that came twelve different families of languages.[155]

And though this modern myth I've just given you ought not to be taken literally, it does weave together a more comprehensive understanding than any tale from the ancient world of why *this moment* looks as it does *now*.

Like so, what this book points to accounts for everything—from consciousness to the cosmos, from Knowledge of good and evil to Knowledge of the two truths, and more.

APPENDIX D:
QUAGMIRES, LACUNAS, AND LONGSTANDING FEQS

As we've seen, the two truths are expressed in patterns of symmetry. One aspect of Reality appears as change, while the other appears unchanging. This can be *seen* and experienced directly. And it **needs** to be *seen*—not believed, speculated upon, or figured out.

Once it **is** *seen*, all the great, unresolved problems, questions, and mysteries of science, religion, philosophy, and life evaporate.

There are countless such "mysteries," "issues," and flat-earth questions that we could examine. This appendix points to a few of them.

As you will see, the two truths permeate anything and everything—including "thingness" and "permeation."

If, after reading this appendix, you have additional concerns or questions, please go to mtsrmts.com, where you can submit them to me.

Let's begin with some longstanding problems of philosophy:

1. *The problem of induction*, or the inferring of general principles from the observations of particulars. The example often cited is to conclude that all swans are white since, supposedly, no one has ever reported seeing a swan that was **not** white. Such an inference held strong among Western people at one time, but eventually Europeans came upon Australia and discovered black swans living there.

This is the problem with induction. It holds only so long as no countering evidence is ever produced. But since, in countless instances, countering evidence can arrive at any moment, inductive reasoning is ultimately not reliable. One can never know in advance what unprecedented particularity might come their way.

This is a problem, however, only so long as we believe that there **are** particulars—i.e., substantial things that persist. Once it is *seen* that we can't say that there really **are** any particulars (e.g., particles, quarks, cats, swans, you, me, or God), the problem of induction vanishes. And so do most other lingering problems of philosophy, as you'll find in, say, E. G. Moore's classic *Some Main Problems of Philosophy*.

2. If you're not sure that virtually all longstanding problems of philosophy are settled with a proper understanding of the two truths, here's another all-consuming conundrum: *free will—do we have it or not?*

Yet there's a contradiction built into the term *free will.* Will is necessarily conceptual. It's bound up in the illusion of self. It's a thing ostensibly owned by, or contained in, another thing. It involves imagining permanence where there isn't any. And it always involves mental grasping.

Since we don't find thingness or permanence anywhere within the conceptualized world, we can't say that there are any "beings"—whether "God" or "us"—that can have or exercise will in the first place. We can't say that there is any will to be had or experienced, either.

This turns out to be an FEQ.

True Freedom never involves will. It's of the perceptual—the Mind of Totality. It's avoiding partiality and living without preference.

3. Remember Samuel Johnson? When his friend Thomas Boswell insisted that it's impossible to refute *Bishop Berkeley's idea that "everything in the universe is ideal,"* Johnson famously kicked a large stone, hurting his foot, and declared, "I refute it thus."

The two truths, however, show that particulars are mere appearances. These include all formed entities such as "stones," "feet," and whatever it is we think we're referring to as "Samuel Johnson."

And then there are conundrums of math and science:

4. Albert Einstein wrote in his essay "Geometry and Experience":

> *How can it be that mathematics,* being after all a product of human thought which is independent of experience, *is so admirably appropriate to the objects of reality?* Is human reason, then, without experience, merely by taking thought, able to fathom the properties of real things?

Again, we can see how a proper understanding of the two truths undoes this question (which, as Einstein noted in the same essay, "in all ages has agitated inquiring minds").

We have *seen* that what Einstein, and just about everyone else, calls "real things" are not actually Real, but merely mindstuff—conceptualized, quantized manifestations of Mind.

In not *seeing* past this fundamental confusion, inquiring minds remain mystified as to why mathematics is "so admirably appropriate to [such] objects."

As we've *seen* repeatedly, the stuff of the World and the stuff of Mind are one and the same. Of course they would seem to be "admirably appropriate."

Calculus is effective in describing motion for the same reason.[156]

This, too, turns out to be an FEQ. It's somewhat analogous to asking, *Why are cats so cat-like?*

5. *Number, also,* no less than the world of form, time, space, and movement, *is mental construct.* Indeed, number exhibits all the contradictoriness of all other mind objects.

This is why, for example, the number 1 appears to equal 0.999 . . . . (The ellipsis means that the 9's continue without end.) Here again we have the two truths: the distinct, conceivable number 1 which equals the inconceivably long 0.999 . . . , which comes to no end.

If you don't trust that 1 = 0.999 . . . , then simply substitute x for 1, as in $x = 0.999$ . . . . All you need to do now is move the decimal point one place to the right (i.e., multiply by the number 10) on each side of the equation, then subtract the original terms from each side and divide by 9. In other words:

1. Let: $x = 0.999$ . . .
2. Multiply both sides by 10: $10x = 9.999$ . . .

3. Subtract original terms from each side:

$9x = 9.000\ldots$

4. Divide each side by 9: $x = 1.000\ldots$

If you're still not convinced, then consider that
⅓, expressed in decimal form, is 0.333 . . . , and ⅔ is
0.666 . . . , then ⅗, which equals 1, is 0.999 . . . .

Like all other objects of the phenomenal world, numbers are strictly manifestations of Mind.

This is why math works when applied to the "real
world."

6. The usefulness of numbers is not limited to describing the world of conceptualizable forms. There are also
numbers that allow mathematics to reach beyond mind
objects.

Quaternions are complex numbers that point to
aspects of Reality where consciousness cannot draw
distinctions.[157]

The square root of –1 (or *i*) provides a simple example.
Unlike "real" numbers that can clearly refer to conceivable objects—"real things," as Einstein put it in item 4,
above—whatever *i* supposedly refers to is not conceivable. That's why people called it "imaginary" (or *i*) when
they first came upon this number. But this number has
proven quite useful in describing many phenomena that
do show up in the so-called "real world"—i.e., the conceptualized world of this and that.

For example, in dealing with certain issues in electrical engineering, say, or in quantum mechanics, the number i often comes in handy. This is so even though, while complex numbers are in use during the calculations, no one is able to visualize or conceptualize what's going on.

It's only when complex numbers have been canceled out, and we're back to dealing with "real" numbers alone, that we can, once again, visualize—and, hence, observe—the "objects" "out there" in the "real world" that we're dealing with.

In other instances, such as in various string theories, the use of quaternions suggests that Reality involves considerably more spatial dimensions than the three we take for granted.

And so, quite naturally, the FEQ comes up: where are these dimensions?

But, as we saw in the main text, we can't even substantiate the three spatial dimensions we're familiar with, let alone other, "extra" dimensions. This is because all dimensions are conceptual rather than Real. So how substantial can we possibly expect these inconceivable dimensions to be?

Currently, the favored explanation of why we don't see these extra dimensions is that they must be "curled up." But where? In some cosmic, multidimensional sock drawer? (Again, the knee-jerk assumption is that dimensions **are** Something, as opposed to concepts.)

The reason we can't visualize more than three spatial dimensions is that consciousness simply isn't capable of the task. If you need proof, just try it. (No gimmicks, tricks, or analogies allowed.)

So, regarding dimensions, we can't say that there are any. Can't say that there are not, either. We're back to the two truths.

7. *The "coincidence problem"* is derived from the observation that the Universe seems to be incredibly fine-tuned to support life.[158] It nicely corresponds to the question of why mathematics seems so well-tuned to the workings of the human mind, as well as to the workings of nature, as we considered in item 4 of this appendix. The stuff of the World and the stuff of the Mind are one and the same. Of course "they" would appear to be in sync.

Some folks hold up this observation as proof that the universe was designed just for us by some creator. Others, for no other reason than to defend their substantialist belief, posit the extreme material extravagance of the many-worlds hypothesis, which requires us to take an awful lot on faith.[159] Still others resort to other hypotheses to explain this "coincidence" away. But all these explanations rely on conceptual constructs and thus miss the mark.

All that is necessary is that we not lose sight of the two truths, and of symmetry. When we *see* that Mind is antecedent to matter, the observation of a finely tuned universe is not a conundrum, a mystery, or a surprise.

8. At the very beginning of this book, in the introduction, I posed the question *"What constitutes a measurement?"* and noted that physicists haven't a clue.

This is particularly true when it comes to quantum systems. This is because, as physicist Rodney Brooks notes in part 2 of Appendix B, field physicists do not understand field collapse or consciousness.

"What constitutes a measuring device?" is an important question for quantum scientists because it has always seemed impossible to pin down just exactly when and where a measurement takes place. For example, when and where does field collapse occur? And what triggered it?

After reading this book, and particularly part 2 of Appendix B, we can see the answers:

What is a measurement? Answer: conceptualization.

What constitutes a measuring device? Answer: consciousness.

Of course, since consciousness is not a thing and nothing produces it, no "device" is ever needed or involved.

And as for religion:

9. There are two ways to live one's life. One is to be preoccupied with the inescapable conceptual fracturing

of the World. The other is to regularly acknowledge this apparent fracturing, while never losing *sight* of Totality.

Religion's true calling lies with the latter—but only as long as it relies on immediate, direct experience and perception, rather than on beliefs and concepts.

Once you realize—with or without religion—that Truth never appears as a concept, you're suddenly in a good position to *wake up*. Having cut through all quagmires and confusion, you're free to trust direct perception. To *know* Reality. To *just see*.

As this occurs, all the vexing and perplexing questions of religion reveal themselves as FEQs.

In order of appearance.

Huang-Po, quoted on pages 5, 25, 68, 129 (twice), 152, 167, and 289, lived 720–840. He was a highly influential Zen master during the Tang Dynasty in China.

Andrew Grant, quoted on page 35, is a science writer for *Science News* and other publications.

Sir Arthur Eddington, quoted on page 47, lived 1882–1944. He was an English astronomer, physicist, and mathematician. He observed the 1919 solar eclipse that provided one of the earliest confirmations of Albert Einstein's theory of General Relativity.

Richard Feynman, quoted on page 98, lived 1918–1988. He was an American theoretical physicist known for his work in quantum mechanics. He received the Nobel Prize in Physics in 1965.

Hui-neng, quoted on page 135, lived 638–713. He is considered the Sixth Patriarch of Ch'an (Zen) in China. All of the several extant schools of Zen can be traced back to Hui-neng.

Dogen Zenji (a.k.a. Eihei Dogen and Dogen Kigen), quoted on page 145, lived 1200–1253. He was a Japanese Buddhist monk, writer, and poet who brought Soto Zen from China to Japan.

Joseph Campbell, quoted on page 149, lived 1904–1987. He was an American professor of literature. He is best known for his book *The Hero with a Thousand Faces* (1949), which presents his theory of the archetypal hero found in world mythologies.

Ikkyu, quoted on page 149, lived 1394–1481. He was an eccentric, iconoclastic Japanese Zen monk and poet. He had a great impact on Japanese art and literature and was largely responsible for infusing Zen into it.

Susan B. Anthony, quoted on page 155, lived 1820–1906. She was an American social reformer and women's rights activist who played a pivotal role in the women's suffrage movement.

Dainin Katagiri, quoted on pages 165 and 201, lived 1928–1990. He was the founder of the Minnesota Zen Meditation Center in Minneapolis. He was my primary Dharma teacher.

Emily Dickinson, quoted on page 169, lived 1830–1886. She was an American poet of the first order.

Nagarjuna, quoted on page 189, lived c. 150–c. 250. He is widely considered to be one of the most important Buddhist philosophers. He founded the Madhyamaka school, and is credited with developing the philosophy of the Wisdom (*Prajñaparamita*) literature of the Mahayana.

Henry David Thoreau, quoted on page 193, 1817–1862. He was an American essayist, poet, philosopher, abolitionist, naturalist, tax resister, surveyor, historian, and leading transcendentalist. Best known for his book *Walden*.

Understanding and Using
a Vocabulary of Enlightenment

In this book, I use some terms and concepts in unorthodox ways. This glossary is a brief guide to these usages.

Absolute There is no dictionary definition of this term when capitalized. In lowercase, however, some dictionary definitions are: not qualified or diminished in any way; total; existing (or experienced) independently and not in relation to other things; not relative or comparative; a value or principle that is regarded as universally valid, or that may be viewed without relation to other things. As used in this book: not relative, not conceivable, not contingent, not dependent, not contrastable, not changeable, not graspable, not objective, not subjective, not formed, and so on.

awareness Some dictionary definitions: knowledge or perception of a situation or fact; feeling, experiencing, or noticing something, such as a sound, sensation, or emotion; concern about and well-informed interest in a particular situation or development. As used in this book: noticing a conceptualized entity via sensation, thought, and/or emotion.

Awareness There is no dictionary definition of this term when capitalized. As used in this book: perception; Knowledge; Objectless Awareness.

belief Some dictionary definitions: an acceptance that a statement is true or that something exists; a religious conviction. As used in this book: a **held** view—that is, any view that is taken or presented as True or Real. Compare with *faith*.

concept Some dictionary definitions: an abstract idea; a general notion; a plan or intention. As used in this book: any formed and individuated thing, thought, idea, feeling, or emotion that is taken to be distiguishable and distinct from any other individuated entity.

conception Some dictionary definitions: the way in which something is regarded or perceived; a general notion; an abstract idea; a plan or intention; understanding; an ability to imagine. As used in this book:

the manifestation of individuated mind objects as the functioning of consciousness. It's regularly mistaken for perception.

consciousness I offer no dictionary definitions, since they tend to be based on flat-earth assumptions. As used in this book: an aspect of Mind that divides Totality. The manifestations of Mind (objects and distinctions) that consequently appear, lack substantiality. Consciousness is the apparent dividing of a seamless Whole into "the world we see"—i.e., the everyday world filled with "this's" and "thats." Field physicists call this division "field collapse."

delusion Dictionary definition: an idiosyncratic belief that is contradicted by what is generally accepted as reality. As used in this book: belief in the substantiality of concepts—i.e., manifestations of Mind.

descry (*dee scry'*) As used in this book: to catch sight of something difficult to catch sight of. Do not confuse *descry* with *decry* (to disparage).

empty There is no dictionary definition of this term when italicized. As used in this book: lacking substantiality.

existence Dictionary definition: having objective reality. As used in this book: persistence.

faith Some dictionary definitions: complete trust or confidence in someone or something; strong belief in God or religious doctrine, without proof. As used in this book: trust; confidence. (People often mistakenly conflate faith with belief.)

form Some dictionary definitions: the visible shape or configuration of something; arrangement and style in literary or musical composition; the essential nature of a thing. As used in this book: any object (or group of objects), such as one or more tables, birds, gods, people, stars, clouds, atoms, ideas, emotions, feelings, concepts, and so on, that appears within experience via consciousness. In all cases, form—whether mental or physical—is illusory. That is, there is never any substantiality "within" or "behind" it.

here There is no dictionary definition of this term when italicized. As used in this book: without location. Not to be contrasted with "there," which appears only within the context of space or locality. Both "here" and "there" are concepts; *here* is not. *Here* alludes to objectless Reality. Compare with *now*.

illusion Dictionary definition: something that is false or not real, but that seems to be true or real; an idea based on something that is not true. As used in this book: any conceptual form(s), whether mental or physical.

know Dictionary definition: to be aware through observation, inquiry, or information. As used in this book: to believe. To trust in the veracity of some particular concept or thing.

know There is no dictionary definition of this term when italicized. As used in this book: to directly realize Totality.

knowledge Some dictionary definitions: facts; information; skills; awareness or familiarity gained by experience. As used in this book: belief.

Knowledge There is no dictionary definition of this term when capitalized. As used in this book: perception; Awareness of Reality; Objectless Awareness.

manifestation(s) of Mind There is no dictionary definition of this phrase. As used in this book: the appearance of any and all phenomena—whether physical or mental, thought or thing. Compare with *mind object*; also compare with *phenomena*.

matter Some dictionary definitions: physical substance, as distinct from mind and spirit; that which occupies space and possesses rest mass, as distinct from energy. As used in this book: physical objects (and/or the stuff they appear to be made of). While commonly thought to be distinguishable from mind, matter is actually Mind Itself.

mind object There is no dictionary definition of this term. As used in this book: any concept; any formed and individuated thing, thought, idea, feeling, or emotion. Any mind object necessarily reveals the subject/object split common to all conscious experience. Compare with *manifestation(s) of Mind*; also compare with *concept* and *phenomena*.

Mind I offer no dictionary definitions of this word in uncapitalized form because such definitions are based on flat-earth assumptions. There is no dictionary definition of this term when capitalized. As used in this book: Mind (with a capital M) does not refer to "your mind" or "my mind," or to anything like our concepts of either. In varying contexts, other words for Mind might be Reality, Truth, Wholeness, Totality, Awareness, Knowledge, The Whole, Thus. Mind is directly *known*. It is sometimes referred to as "the One Mind" since, as it is possible to *see*, there is **only** Mind.

Nothing There is no dictionary definition of this term when capitalized. Nor could there be.

now There is no dictionary definition of this term when italicized. As used in this book: without time. Not to be contrasted with "then," or confused with "the present moment," which appear only within the context of time. Both "now" and "then" are concepts; *now* is not. *Now* alludes to objectless Reality. Compare with *here*.

perception Some dictionary definitions: the ability to see, hear, or become aware of something through the senses. As used in this book: Objectless Awareness; Knowledge; *just seeing, just hearing, just smelling, just tasting, just touching.*

phenomena Dictionary definition: the objects of sensory perception. As used in this book: manifestations of Mind that are regularly mistaken as objective, Real, and substantial.

real Some dictionary definitions: actually existing as a thing or occurring as a fact; not imagined. As used in this book: conceptual experience; manifestations of Mind that only **appear** to exist as things, or occur as facts.

Real There is no dictionary definition of this term when capitalized. As used in this book: perceptual experience.

reality Some dictionary definitions: the world or the state of things as they actually exist, as opposed to an idealistic or imaginary idea of them; a thing that is actually experienced or seen; a thing that exists in fact; the state or quality of having existence or substance. As used in this book: the phenomenal world—the world of apparent time and space, mind and body, physical forms, mental impressions, thoughts, feelings, and sensations. **All such phenomena are illusory**.

Reality There is no dictionary definition of this term when capitalized. As used in this book: Truth. Mind. The Whole. Totality. Reality isn't anything in particular—nor can It be. It's simply Thus.

seeing Some dictionary definitions: perceiving with the eyes; discerning visually; becoming aware of something from observation, or from a visual source. (These definitions all conflate perception and conception.) As used in this book: visual conception.

seeing There is no dictionary definition of this term when italicized. As used in this book: pure perception, which is not limited to (or located in) eyes and brains. *Seeing* is immediate and without limits or location.

self Some dictionary definitions: a person's essential being, which distinguishes them from others; the object of introspection or reflexive action. As used in this book: a purely conceptual construct; an illusory object (or subject) of the phenomenal world that appears to persist even as "it" supposedly changes and moves about (hence, the illusion). A self would have to be something that doesn't change, but we never find such entities within perceptual experience. We only imagine and believe that we do. Compare with *Real*.

something Dictionary definition: a thing that is unspecified or unknown. As used in this book: the mistaken view that phenomenal appearances are substantial; a mind object.

Something There is no dictionary definition of this term when capitalized. As used in this book: the mistaken view that phenomena are substantial.

substantiality Dictionary definition: the quality of being real and tangible rather than imaginary. As used in this book: the Realness that we imagine and (mistakenly) grant to phenomenal appearances. Such appearances are referred to as *nonsubstantial* in this book rather than as *insubstantial* to avoid any implication of existence.

symmetry Some dictionary definitions: the quality of being made up of similar parts that face each other (bilateral symmetry), or that surround a central axis (radial symmetry); similarity or exact correspondence between or among different things. Physicists' definition: any transformation within a system that leaves the overall system unchanged. As used in this book: even as phenomenal forms appear in constant flux, Reality remains forever Thus. Even as you seem to persist, all the elements of your mind and body appear to arise and pass away. Even as everything in the Universe manifests as thoroughgoing change, in Totality the Universe appears as without intrinsic size, duration, location, or substance. Thus, the World everywhere and at every turn reveals symmetry between two truths: relative truth and Absolute Truth, or simply truth and Truth.

then Dictionary definition: a previous or other time. As used in this book: in Reality, there **is** no "then." "Then" is a concept, an illusion. Compare with *now*.

there Dictionary definition: a different or other place. As used in this book: ultimately, there **is** no "there." "There" is a concept, an illusion. See *here*.

This There is no dictionary definition when capitalized. As used in this book: what is actually *seen*—i.e., Truth, Thus, Reality.

Thus There is no dictionary definition when capitalized. As used in this book: Reality (with a capital R). Totality. The Whole. The Knowledge that **This**, though nothing in particular, is not Nothing.

Totality There is no dictionary definition when capitalized. As used in this book: The Whole—which is not an object. Reality. Truth. Thus. The One Mind.

truth Dictionary definition: the quality or state of being true. As used in this book with lowercase initial: relative truth. Relative truths correspond to "facts" about the illusory phenomena of the relative world.

Truth There is no dictionary definition when capitalized. As used in this book: Reality (with a capital R). Totality. Awareness. Mind. Knowledge. Wholeness. Thus.

two truths There is no dictionary definition of this term. There are two truths: relative and Absolute. Consciousness is the conceptualization of relative truths; Awareness is the perception of Absolute Truth. The two truths are neither separate nor the same. This is Ultimate Truth.

unwholesome Dictionary definition: Not conducive to physical or moral well-being. As used in this book:

partial; not concerned with or attentive to Wholeness, or to the Whole.

wake up There is no dictionary definition of this phrase without a hyphen. As used in this book: to break free of words and concepts; to realize immediate, direct experience.

Whole There is no dictionary definition when capitalized. As used in this book: total; complete; nothing left out, excluded, or omitted.

Whole, The There is no dictionary definition for this term. As used in this book: Totality. Reality. Mind.

Wholeness There is no dictionary definition when capitalized. As used in this book: Awareness (with a capital A). Truth. Reality. Totality. The One Mind.

Wholesome There is no dictionary definition when capitalized. As used in this book: concerned with or attentive to Wholeness, or to the Whole.

world, the Dictionary definition: the Earth, together with all its countries, peoples, societies, and natural features; the material universe or all that exists.

World, The There is no dictionary definition when capitalized. As used in this book: Totality.

NOTES

1 Although it's my hope that, in most cases, ANYONE will speak for you, I've created a website, mtsrmts.com, where you can raise questions or issues that you feel are not adequately dealt with in this book.

2 Many readers of my previous books tell me that they reread them multiple times. If you are such a reader, I encourage you to, perhaps, read through the main text once, without generally referring to these notes, and then to reread both the text and the notes at your leisure. The notes provide a great deal of additional range and depth to what appears in the main text.

3 For a more thorough understanding of the distinctive ways the terms *truth* and *Truth* are used in this book, consult the glossary and appendix A.

4 You can listen to *The Unanswered Question* by Charles Ives by visiting my website at mtsrmts.com.

5 Gottfried Wilhelm von Leibniz, "The Principles of Nature and of Grace, Based on Reason."

6 In 1803, Thomas Young demonstrated that light appears wave-like, an observation that was bolstered by the work of James Clerk Maxwell a few decades later. Yet in 1905, without disproving Young's experimental results, or quibbling with Maxwell's equations, Einstein used the photoelectric effect to demonstrate that light seems particle-like. This was our first inkling of light's apparent dual, wave/particle nature.

7 These are actual historical arguments that were made against the notion of the "Antipodes" (a place where people's feet were opposite) by Saint Augustine, Lactantius ("the Christian Cicero," chosen by Constantine to tutor his son), Chrysostom (patriarch of Constantinople), and many others.

8 Many of these FEQs may strike you, respected reader, as being perfectly legitimate at first. As you move further into this book, however, more and more of these questions may start to reveal themselves as at least somewhat misguided. I encourage you bear with me!

9 Actually, in quantum field theory (as opposed to the standard model of quantum theory), there are no particles, only fields. For physicists who subscribe to quantum field theory, the enigmas of quantum mechanics don't exist. Instead, they have other, more subtle problems. We'll get to those later.

10 This is but one example of many we will encounter in which we neither find substance nor the lack of it. Much more on this in the pages to come.

11 For more detail, read "What Is Real?" by Meinard Kuhlmann, *Scientific American*, August 2013, pp. 40–47. These examples are noted on page 43 of that issue.

12 Actually, if there **were** Something substantial, we would **not** be able to "build upon It." As we shall see, the very appearance that we **can** build upon things directly indicates that "they"—these "somethings"—are, in Reality, **non**substantial.

13 As we'll see, it's like trying to progress along a Koch curve—i.e., an infinite line within a finite space.

14 If you're interested in physics: the "collapse" of the Schrödinger wave function coincides with a non-manifest "probability wave" (as particle physicists would have it) suddenly appearing as a physical entity—an electron, say. Schrödinger himself, however, did not interpret his own equation this way. He saw it as a description of the inten-

sity of an oscillating field, manifesting itself as interfering waves. According to his description, the particle-like behavior of the field is the collapse of the field, no matter how spread out it may have been, as it deposits all its energy into a single point. It's *just this* with no need to conjure up schizoid "point particles." In other words, to field physicists, of whom Schrödinger was a forerunner, there **are** no particles. The thing that remains a mystery for field physicists, however, is field collapse itself. Apart from an oscillating field depositing all its energy into a single point (and what, exactly, is that?), how does it come about? Both field physicists and particle physicists haven't a clue. This is because they profoundly misunderstand consciousness, as we'll see. (You can find more on field collapse in part 2 of appendix B.)

15 The reference here is to famous experiments by Benjamin Libet et al. What neuroscientists don't understand in regard to Libet's results is that it is not "you," the subject, who prepares to move. Much more on this later.

16 Analysis of twelve-billion-year-old light shows signs that the *fine structure constant*, more commonly known as *alpha*, was of a slightly different value 12 billion years ago. This observed fact, if true, has enormous implications for physics, since it means we now operate from a model of the universe that is in error. See *Thirteen Things That Don't Make Sense* by Michael Brooks, pp. 46–56.

17 You may think you don't *know* Truth and Reality, but this is impossible. Truth and Reality are all you've ever experienced. The problem has to do with what you think—more precisely, with what you believe. More on this to come.

18 Adherents of this basic belief are also called *physicalists*, and they come in a variety of flavors.

19 Materialism is, basically, the philosophical belief that everything that exists is ultimately material. There are varieties of materialists. **Reductive materialists**, for example, believe that everything reduces to the physical. For them, men-

tal states are nothing more than brain states. ***Eliminative materialists***, on the other hand, believe that eventually our notions of mental states—fears, joys, hopes—will be eliminated once the science of human brain functions advances far enough. According to this latter view, mental states will be rejected in much the same way that concepts such as humors in the blood or demon possession were eliminated once better theories came along. As we'll explore in this book, however, there are major, inescapable, and insurmountable flaws intrinsic to all possible forms of materialism. Like all beliefs and philosophical theories, materialism is ultimately untenable, no matter its configuration.

20 Neuroscience researchers such as Antonio Damasio, et al. define ***mind*** as the flow of mental images. Quite apart from the many problems that arise with the word *flow*, this definition assumes that this flow of images takes place inside the brain. This is still a materialistic interpretation or explanation of mind—as if mind were materially based and somehow constructed by the brain.

21 Aside from helium and neon—two inert gasses not given to combining with any of the other naturally occurring elements—hydrogen, oxygen, nitrogen, and carbon, the elements essential to life, are the most abundant elements throughout the universe.

22 This is not to say that we can't *know* life, mind, or consciousness. As we shall see, we **already** *know*.

23 If you're interested in cosmology: There is now speculation that newly discovered "primordial black holes"—that is, black holes formed not from collapsing stars, but from dense regions of the primordial universe—could account for the missing mass in the universe. This opposes the idea that undetected particles are needed to account for the missing mass. But even if this more recent conjecture is found to be the case, it will still not help us see what matter actually is. This ignorance will remain with us for as long as we continue to think that matter actually **is** Something.

24 This, of course, is matter and energy from a materialist's point of view, which explains or reveals nothing.

25 From "Matter Gone Missing" by Andrew Grant in *Discover*, June 2010, p. 16. For all we know, there **is** only ordinary matter, but most of it is undetectable. There is evidence that suggests this. Of course, as ANYONE optimistically suggests, science will work it out—which may very well be true. But it will also generate ever more questions in the process.

26 Dark energy apparently makes the universe fall outward. In other words, its expansion is accelerating, as though it were in free fall.

27 If you're interested in physics and cosmology: For more, see *A Universe from Nothing*, by Lawrence M. Krauss. For example, on page 98, Krauss writes, ". . . if inflation is . . . responsible for all the small fluctuations in the density of matter and radiation that would later result in the gravitational collapse of matter into galaxies and stars and planets and people, then it can be truly said that we all are here today because of quantum fluctuations in what is essentially *nothing*." He adds, "Quantum fluctuations, which otherwise would have been completely invisible, get frozen by inflation and emerge afterward as density fluctuations that produce everything we can see! If we are all stardust . . . it is also true, if inflation happened, that we all, literally, emerged from quantum nothingness [i.e., from quantum fields]." This "nothing" that Krauss speaks of here, however, is not Nothing. At a minimum, there is at least the Higgs field.

In addition to these considerations, there are further implications regarding the lack of substantiality of particulate matter that can be drawn from quantum field theory, as we shall see.

28 Positive energy appears in the form of matter—the energy released in nuclear explosions, for example. Negative energy appears in the form of gravity.

29 Were the tabletop expanded in this manner, it would be about ⅝ of a mile on each side.

30 Were the actual tabletop expanded by this amount, it would now be a little more than 621 miles long on each side. That is greater than the distance from New York City to the South Carolina border.

31 Were the original tabletop expanded by this amount, it would be 6,213,712 miles on each side, reaching far past the orbit of the moon, to nearly 7 percent of the distance to the sun.

32 If you're interested in physics: We're talking about the look and feel of things. What you feel and see as a solid surface is just these pushes and pulls, as with two bar magnets. In other words, what you see as color and surface are neither color nor surface belonging to an object. This extends to other sensory experiences as well. For example, the watery cool feeling of freshly sifted, finely ground flour, as with the feel of water itself, is just the apparent tininess of the loosely bound water molecules or starchy granules. The same also occurs with the senses of smell and taste. Likewise heat and sound. None of these experiences turns out to be what we naïvely think it is.

33 And this applies to what we naïvely take as the "surface" of a billiard ball.

34 If you're interested in physics: You may have seen, for example, the famous image on a computer or TV screen in which IBM spelled out its initials in atoms. You'd be forgiven if you thought you were looking at individual atoms, well defined within their spatial surfaces, like tiny welts on some sort of undefined plate. But these images are not of actual atoms; they are generated electronically from readings of where the pushes and pulls are positioned in relation to each other. And notice that the images of the relative locations of the "atoms" seem to appear as if imbedded in a flat surface. What do you suppose that is, other than an electronically generated image? The idea

that we've taken pictures of actual, well-defined, physical objects is nothing more than our own mental construction. Where is the "solid" object in all this? You'll not find it. And don't forget that all we've done here is make use of a "magnifying glass," so to speak, so we can get in close for a good look. Nothing has otherwise been altered.

35 Physicists call "it" a *point particle* because they're all but forced to reach for a reified image—i.e., they feel compelled to put a "something" "there." It's like reverting to a bad habit. What you need to remember is that a "point" has zero dimensions. It's not a thing extended in space. It's a location, but no thing occupies it.

36 If you're interested in physics: At 10^{-15} meters, which is where we are now, we're inside the atomic nucleus, among the high-speed tumbling and vibrating that comprise the roiling realm of quarks. Here their movements are of such velocities that, given the relativistic effects, time ticks off more slowly for quarks—or so it appears from our perspective.

And how do we discover these quarks? Not by physically looking (we've long since left such quaint procedures behind), but through smashing these subatomic particles together in gigantic machines—and then, with supercomputers, analyzing the streams of data that pour forth. For months, such round-the-clock analytical computations go on until we accumulate enough processed information, out of reams upon reams of otherwise discarded data, to justify holding a debate among physicists in which we might decide that, yes, indeed, we discovered yet another quark.

37 As we shall see, the fact that there is no real size involved in this observation is indicative of Mind, not matter.

Given all this, and much, much more, we might want to (once again) question whether there are any particles at all. Just because the ancient Greeks steered us onto the idea, and just because the idea is so graspable, it doesn't mean there's any substance behind the notion.

38 If you're interested in philosophy and mathematics: There are five paradoxes attributed to Zeno of Elea, the first four dealing with motion. Of these, the first two have been successfully dismissed by later thinkers. It's his third and fourth paradoxes that still give people trouble. There are only three possible ways to escape these paradoxes. The first would be if space and time are not corpuscular—that is, if they're not made of discrete intervals or points. The second would be if space and time **do** have moments and points, but these moments and points are not only infinite in number, but **also** infinite in number even within any given interval of time or space (such as within a millisecond or a micron), no matter how small. The last escape is that space and time are illusory.

Most mathematicians, and probably most scientists, now assume the second option is true, since the first doesn't hold up very well according to their analysis. As for the third option, they regard it as a joke—*even though this is the conclusion Zeno wanted us to come to.* But it's not just option one that doesn't hold up; option two doesn't hold up either, as we'll see. Only with this third case does everything become clear. More on this later.

39 Though this example has been widely noted, Richard Dawkins nicely captured it in a *Free Inquiry* article, February/March 2009, pp. 28–29. I am relying on his observations here. It is interesting, though, that Dawkins grants that intelligent design actually explains a few things. In fact, it explains virtually nothing, while assuming a great deal without supporting evidence. Dawkins also asks, "If any reader knows of an idea that has a larger explanation ratio than Darwin's, let's hear it." In response, I'd offer Dawkins this book—except that I'm not explaining anything here, and not proposing ideas. I'm only pointing things out.

40 Some people do say this. *Metaphysical nihilists*, as they are sometimes called, are philosophers who believe in the possibility of Nothing. Most champion the *subtraction*

argument, which claims that Nothing can be arrived at by simply subtracting everything—provided we accept the premises that: (1) the world contains a finite number of things; (2) these things are all contingent, meaning that they might **not** have existed; and (3) the nonexistence of any one thing does not entail the existence of some other thing. As we will see, each of these premises conflicts with Reality.

There are additional arguments for, and definitions of, Nothingness as well. Imagine, for example, post-Einstein, that we live in a closed, curved spacetime analogous to, say, the surface of a balloon. Now, imagine the balloon shrinking to a radius of zero. "This thought experiment leads to an elegant scientific definition (originally due to the physicist Alex Vilenkin): Nothingness = a closed spherical spacetime of zero radius." Jim Holt, *Why Does the World Exist?*, p. 50. Incidentally, I highly recommend Holt's book to anyone interested in the subject.

But slow down and reconsider Vilenkin's observation: Nothingness = a closed spherical spacetime of zero radius? No, it doesn't. How can Nothingness be equivalent to anything? This is a meaningless equation, loaded with flat-earth assumptions. For example, it assumes an objective reality of spacetime. It also ignores the Higgs field that, physicists tell us, fills all of space. Even in empty space, when all other fields are at zero, the Higgs field is not at zero. And, in fact, it can never go to zero. So, we never have Nothing. Metaphysical nihilists neglect to take this, as well as any number of other things, into account.

41 To the extent that our thinking does **not** follow these laws, we will **not** be able to make conceptual sense of the world. For example, if you try to have a conversation with someone who didn't obey the law of contradiction, your conversation would break down in very short order:

"I heard you just got back from France."
"Yes, and we had a wonderful time."

"Oh, I know. I just love strolling the Champs-Élysées this time of year."

"When were you last in Paris?"

"Me? Never been there."

"You just told me how you love strolling the Champs-Élysées!"

"No, I've never been to France."

"Then how could you have strolled . . ."

"Did you get a chance to tour the Louvre? I managed to see the Mona Lisa on my last visit, though I must say I was startled by its size. I always imagined it bigger."

Whether we're aware of it or not, we need to think and speak in accordance with these laws, or we will not be able to communicate in any coherent way. This gives us the strong impression that Reality, too, must comply with these laws. But, as we shall see, this is not at all the case.

42 To say that a proposition must be either true or false is to say that it must be either provable or disprovable by the laws of logic. Gödel, however, demonstrated that there are propositions that are neither provable nor disprovable. He did this by setting up a mathematically valid statement that said, in effect, "This statement cannot be proven." Once there, the rest comes easily. If you can prove it, then it's not true. If it's true, then you can't prove it. For more, see Morris Kline's *Mathematics: The Loss of Certainty*, p. 264.

This has many implications that might not be immediately obvious—including, to take just one example, why concepts don't and can't yield Ultimate Truth.

This will become more obvious as you continue in the main text.

43 Kurt Gödel's incompleteness theorem didn't actually shore up quantum mechanics (which all but fades from view in light of quantum field theory), so much as it provided evidence for a better understanding of consciousness than our current theories can supply. In turn, this better understanding of consciousness sheds light on why mathematics

applies so well to the world we observe—a question that famously mystified and eluded Albert Einstein and other moderns, as well as the ancient Greeks and earlier thinkers. For more on this, turn to appendix D, item 4.

44 Actually, as we will see, this is True of **all** of what we call "the physical universe."

45 I am capitalizing Something and Nothing here because this is how Leibniz—as well as most of the rest of us— used (and continue to use) these terms. In other words, we believe Something and Nothing are Absolute Truths. For most of us, Leibniz included, it's Absolutely True that there is Something. It's also Absolutely True that there is not Nothing. But we'll soon *see* that these are flat-earth assumptions.

46 Which is precisely how most people unwittingly conceptualize Reality.

47 I'll use the ordinary term "see" when referring to a conceptualized object, and *"see"* (italicized) when referring to perception. *See also seeing* in the Glossary.

48 Wittgenstein once asked a friend, "Why do people always say it's natural for people to assume that the sun goes around the Earth rather than that the Earth is rotating?" "Obviously," countered his friend, "because it **looks** as if the sun is going around the Earth." To which Wittgenstein replied, "Well, what would it have looked like if it had looked as if the Earth was rotating?"

But Wittgenstein was no less caught up in conceptualization than those he questioned—perhaps even more so. After all, it's not as immediately obvious that the Earth is a spherical planet that rotates in relation to the sun.

Like most of us, Wittgenstein had wandered so far down the road of conceptualization that he ignored what he actually *saw* due to the overriding influence of what he thought. He, too, created a conceptual structure and got lost in it, ignoring that there can be no veridical concept to represent, model, or stand in for Reality. Reality (with a big *R*) is not a concept. It can only be *seen* and *known* directly—i.e., perceived, not conceived.

49 Oliver Sacks, *An Anthropologist on Mars*, pp. 108–52. Elsewhere, Sacks famously reminds us of the man who mistook his wife for a hat (the title of another of his books). These and other stories clearly indicate that perception is something quite distinct from conception.

50 Ibid., pp. 114–15.

51 Ibid., p. 119.

52 ANYONE still believes that their mind has a location.

53 I am not pointing this out because "coffee cup" is merely an idea, but because, as we'll *see*, "coffee cup" has no actual perceptual referent at all.

54 The term "after" is in quotes because, as we shall see, time is also a conceptual construct—and, thus, an illusion.

55 Even though we may see, and consequently believe, that there's a tree three feet away from us, this does not mean that we've captured Reality through this observation. In fact, as we will discover, we can be certain that we have not. Indeed, Reality cannot be captured, only *seen* and experienced directly—i.e., nonconceptually.

56 We construct things out of perception all the time in this way. For example, we readily see faces where there aren't any. We see them everywhere—in flowers, in clouds, in tree bark, even on the surface of the moon. Even in a breakfast roll. It has nothing to do with whether or not you put significance on it. Such images seem to pop out at us.

We even see colors that are "not there," because we think we're viewing a "white surface" in shadow and interpret it as quite distinct from the identically colored "gray surface" that we think is in light. (For a stunning example of this, Google "checkerboard illusion.") And if we stare at a white screen arrayed with green dots placed in a circle, when the green dots are momentarily whited out in contiguous sequence, a phantom "violet dot" will appear to move around the circle in their place. (Google "color illusions.")

We can see even more elaborate movement that is "not there." When we go to a theater and watch a film, the movement appears so convincing that we forget we're only viewing flickering streams of still photos.

But such deception is not limited to the visual. We can hear steady tones that appear to change in pitch, or sample apples that taste like pears, or feel both warmth and coolness at once by placing our two hands, one chilled and the other warmed, in the same bucket of lukewarm water.

And, indeed, such deception runs even deeper. In fact, it is thorough. For another stunning example, check out Donald Hoffman's TED talk titled "Do We See Reality as It Is?" In relying on perceptual tricks, or "hacks," as Hoffman calls them, rather than on accurately "perceiving" the world, Hoffman questions our take on reality altogether. Unfortunately, like virtually everyone else, he conflates perception with conception—and, thus, misses the fact that we don't actually perceive objects at all but, rather, **conceive** them. And, as we'll see, this makes all the difference in terms of clarifying our otherwise confused understanding of Mind and consciousness.

57 For more on the "hard problem," see appendix B, part 1.

58 According to *Origins—A Short Etymological Dictionary of Modern English*, by Eric Partridge, p. 663, the word "substance" comes from the Latin **substare**, "to stand under, hence, to resist, to subsist, to persist"—i.e., to exist. It refers to "that which underlies appearance" (ibid.), or to the presumed underlying nature of a thing—i.e., that it's real, it exists, it persists.

Over time, particularly since Descartes, the term has become ever more synonymous with matter. Thus, for us moderns, the basis of reality—what is real, what exists—is primarily considered to be matter. In other words, most of us are basically (and unwittingly) materialists (i.e., substantialists).

59 In fact, the past, like the future, isn't uniquely determined

either, as we shall see. To give a quick example: in his book *The Grand Design*, Stephen Hawking writes, "Quantum physics tells us that no matter how thorough our observation of the present, the (unobserved) past, like the future, is indefinite and exists only as a spectrum of possibilities. The universe, according to quantum physics, has no single past, or history." Stephen Hawking and Leonard Mlodinow, *The Grand Design*, p. 82.

60 As we'll also see, however, thoroughgoing change doesn't mean anything is changing.

61 The term *consciousness* means many things to many different people. On the one extreme, we have people thinking that only humans have consciousness. On the other, we have people who grant consciousness to amoebas. Still others speculate that trees and rocks are sentient, or even molecules and atoms, or stars and galaxies. All this confusion stems from not attending to direct experience—which, as we've noted, does not provide us with particular things that can somehow possess the attribute of consciousness in the first place. For more on Mind and consciousness, please turn to appendix B.

62 As we shall *see*, we are always perceiving Wholeness, never "thingness" or "this-and-thatness." This realization is essential to understanding consciousness.

63 In fact, we can only be fully *aware* when words, labels, explanations, and concepts, are not engaged. This includes the use of concepts such as "we" and "engagement."

64 As with the term *see*, I'll use the term *known* (in italics) whenever referring to perception—i.e., to direct experience.

65 Richard Feynman, *QED*, p. 82.

66 The term Knowledge, with a capital K, refers to objectless Awareness, as opposed to conceptual knowledge—which, as we shall see, is actually belief, and not Knowledge at all.

67 Temporarily, a visual field, or any of the fields described, can appear static rather than ever-moving. But that's simply

because the constant change may, in some cases, be slow, subtle, or so pervasive that it's difficult to discern. At the cellular, molecular, and atomic levels, of course, such change is constant, and neither slow nor subtle. To the extent that we do not recognize static appearances as illusory, we tend to imagine permanence (i.e., existence) where there isn't any—for example, in our concept of "me."

68 At first blush, it might seem easy to argue with what I'm pointing out here. "Of course one color becomes another! Of course the voice of the owl or the robin, or the drip from the faucet, starts and stops!" I understand. I'm well acquainted with these impressions, too. But they're knee-jerk expectations. If you stop your mental chatter (this is imperative) and attend very closely to **actual** experience, you may notice that when it's "on," there **is** no "off;" when it's "off," there **is** no "on." And there is no in-between. It's either "this" and "not that," or it's the reverse. This is actual conscious experience. There is no actual starting or stopping or dividing of anything.

69 These divisions can appear to be spatial, temporal, or both.

70 These two aspects of Mind comprise the subtle nature of Reality. Scientists refer to these dual aspects of Reality as *symmetry*. Understanding their nature is imperative for any understanding of conscious experience. Much more on this to come.

71 This remains the case even when we turn our attention to ourselves—i.e., to the subject "me," thus effectively turning the "subject" into the "object." (Such easy mental manipulation, by the way, is characteristic of mind objects—which, again, reveals their nonsubstantiality.)

72 This doesn't mean that 90 percent of your body mass is bacteria. Bacteria are prokaryotes—that is, their DNA are not enclosed in nuclei like your DNA, but move about freely within each cell's cytoplasm. "Your" cells are eukaryotic and have true nuclei. Eukaryote cells are thousands of times bigger than prokaryote cells, on average.

So, even though 90 percent of the **cells** in "your" body do not carry "your" DNA, the volume of eukaryotic cells in your body is, on average, nearly 900 times the volume occupied by bacteria. Most of these bacteria, by the way, are beneficial.

73 What is doing the raining? No need to answer. Indeed, no need to ask the question. Also notice that there is no confusion and no mystery about raining.

74 This is not a casual observation. It points to symmetry. We will return to this.

75 If everything changes rather than persists, then there can never be anything "there" **to** change. This is the nature of Reality.

76 It is precisely when we **do** believe that objects precede Mind—i.e., that objects are truly objective and "out there"—that we remain at a total loss about consciousness.

77 My friend and colleague Norm Randolph coined the term *unpindownable*, which beautifully draws our attention to a ubiquitous occurrence that usually goes unnoticed by nearly everyone. As Kurt Gödel might have put it, "If you can pin it down, then it can't be True. If it **is** True, then you can't pin it down." This is the nature of Reality.

78 If you're interested in physics: This is also true of quarks and other point particles. They display this same Mind characteristic.

There are many other examples where physical phenomena are better understood as simply characteristics of Mind—such as when, say, in carefully examining changes in velocity, we discover the equivalency of gravity and acceleration.

We'll look at this more closely later on.

79 If you're interested in numbers: When I point this out to people, some come back with the observation that we **do** comprehend such numbers, simply because we can write them down and manipulate them through mathematical

calculations. That's not what I'm pointing out here. I'm talking about Mind and consciousness.

You can easily comprehend at a glance 2, or 3, or even 6 dots on a page, but you'll likely not readily comprehend 87 dots on a page (though some people could). You will probably have to count them in order to arrive at that number—but that number has already become somewhat vague, compared to your immediate grasp of 2 or 3. And with larger numbers still, you might want to count them by learning to see them in groups of five or seven. But already, even with relatively small numbers, your arrival at, say, 537, is considerably less sudden and acute than your immediate comprehension of 3.

Strictly speaking, however, if you examine any number long enough, even the number 1, it will become ever more difficult to comprehend the nature of the object you're considering. At least it will to the extent that you think it actually **is** Something.

More on numbers in item 5 in appendix D.

80 Of course, the photon would have no need for a timepiece. Given that photons are always at light speed, to them, all timepieces appear stopped. If you think that, for them, their own timepiece wouldn't appear stopped, just remember that photons don't carry timepieces. Timepieces are strictly for entities that can't travel at light speed, like us.

This all has to do with understanding consciousness, as we'll see.

81 If you're interested in physics: Regarding time dilation and spatial contraction, Albert Einstein wrote, "The question of whether the Lorentz contraction does or does not exist is confusing. It does not 'really' exist insofar as it does not exist for an observer who moves. It really exists, however, in the sense that it can, as a matter of principle, be demonstrated by a resting observer." The Lorentz-Fitzgerald contraction (its full name) is the reduction in the length

of a body moving relative to an observer, as compared to the length of an identical object at rest in relation to that observer. (The moving object would appear to contract by a factor of $\sqrt{(1-v^2/c^2)}$, where v is the velocity relative to the observer and c is the speed of light.) Other discrepancies can also be witnessed. Time, for example. Clocks and heartbeats belonging to the "mover" appear to tick more slowly as the "mover" approaches the speed of light, as witnessed by the "stationary observer." The "mover," however, observes no such slowing down of their watch or heartbeat.

And remember, either of these participants can consider themselves to be stationary. This means that, as they stare at each other, each sees the other as moving. So, as Einstein mused, do these effects exist, or do they not? Are they Something or Nothing? This will remain confusing only so long as we continue to hold to a substantialist view, and fail to understand the nature of conscious experience.

If you're **really** interested in physics: For an extreme view of time dilation, consider what would happen if you fell into a black hole. If the black hole is large enough, you could actually survive the fall through the event horizon. Yet to anyone observing you from outside the event horizon (i.e., from outside the black hole), you would appear to slow to a standstill, as if you were eternally frozen at the event horizon. You, however, would not experience any slowing of motion or time. Yet, as you pass through the event horizon, from your perspective (if it were possible for you to have one), all of eternity—that is, the entire future of the universe—would have passed by "outside" the event horizon.

Bear in mind that these are only relative views.

The question is: how is all of this in Totality? In other words, what is Real?

Time and space are only for "stationary" entities, like us. It might seem, then, that there must be two sorts of entities: Eternal—i.e., forever *here/now* beings—that never move about or change, and ephemeral entities that never rest.

But in Reality—in Totality—these two are not two. There's only Mind. Stay tuned.

82 This is why physicists have never actually pinned one down, either in time or in space. A photon, like all point particles, is purely a mental construct, not an Object. (Not that any object can be Real. Remember, all objects are manifestations of Mind.)

83 Here is the full quote of Huang-Po, speaking of the One Mind—i.e., Reality, Totality, the Universe: "There is only the One Mind, beside which nothing exists. This Mind, which is without beginning, is unborn and indestructible. It is neither green nor yellow, and has neither form nor appearance. It does not belong to the categories of things that exist or do not exist, nor can it be thought of in terms of new or old. It is neither long nor short, big nor small, for it transcends all limits, measures, names, traces, and comparisons."

84 To quote physicist Nick Herbert from his book *Quantum Reality*, p. 212, "The structure of Bell's proof is as follows. . . . Bell **assumes** that a local reality exists [our commonsense view]. With a bit of arithmetic, he shows that this locality assumption leads directly to a certain inequality (Bell's inequality), which the experimental results must satisfy. Whenever these experiments are done, they violate Bell's inequality. Hence the local-reality assumption is mistaken." Since the publication of Herbert's book in 1985, ever-more-sophisticated experiments have been carried out. All continue to violate Bell's inequality. There have been no exceptions. In short, Reality is non-local. We cannot say there really is a "there" apart from "here." It only appears that way.

85 In other words, these bits of matter formed as a tradeoff with the kinetic energy lost from the original colliding particles.

86 A thing can't move where it is because movement involves taking it to where it isn't. And it can't move where it isn't because it's not "there."

87 It's important to understand that it's not just McLaughlin's interpretation that I will be singling out here. The notion that motion resides in these conjured-up infinitesimals is based on a recent formulation of calculus developed in the 1980s that has since become the standard among scientists and mathematicians. It's also important, however, to note that this most recent formulation follows centuries of handwringing by mathematicians, philosophers, and scientists, based on their frustrated attempts to comprehend the nature of motion. As we will see, our primary confusion over this issue has always stemmed from failing to realize the nature of Mind and consciousness.

88 This statement comes from a famous story in the *Mumonkan*, Case 29, of two Zen monks arguing over a flag they see blowing in the breeze. One says it's the flag that is moving; the other says it's the wind. Hui-Neng, the great teacher of Zen in ancient China, happened by and overheard them. "Not flag," he said. "Not breeze. Mind is moving."

89 Where the original AZ line had a length of 3 (counting 1 for each segment), it now has a length of 4 (i.e., 4 segments). In other words, we've multiplied the original length by $\frac{4}{3}$ (since $\frac{4}{3} \times 3 = \frac{12}{3}$, which reduces to 4). Thus, we've increased the line's length by $\frac{1}{3}$, so that it is now equal to $\frac{4}{3}$ its original length.

90 If, after the first iteration, you multiply the new length, 4, by $\frac{4}{3}$, the second iteration of the line's length will equal $\frac{16}{3}$, which reduces to $5\frac{1}{3}$. In other words, the increase of the second iteration is larger than the increase of the first. Like so, the increase of the third iteration will be larger than that of the second, and so on. With each new iteration, the line increases its length by a greater and greater amount, since you are multiplying $\frac{4}{3}$ by an ever-increasing

number (representing the line's length). Thus, while the first iteration only increased the line's length by 1, the fourth iteration will increase the line's length by nearly 2½. With the seventh iteration, the line increases by more than 5½ times; with the ninth iteration, by nearly 10 times; and with the twelfth iteration, by nearly 24 times. The line's length thus appears to accelerate toward infinity with each additional iteration.

91 We have already noted that consciousness is the mental fragmenting of the World into **apparent** packets of "this" and "that." Now we *see* that such conceptual packets manifest as "motion" and "time" as well. In other words, the fragmenting of the World (i.e., the conceptualization of the World) is not only spatial, but also temporal.

92 If these apparent individuated entities **were** more than mere appearances (in other words, if the objects they purport to present were Real, objective things unto themselves, as we naïvely take them to be), they'd be Absolute—and, hence, as we've noted, incapable of appearing in conscious experience. But since they **do** appear in conscious experience—indeed, they are what conscious experience is comprised of—they are (and must be) appearances **only**.

93 **This** is how Cartesian dualism actually fails. It doesn't fail because the world is basically material, as commonly thought. It fails because there is **only** (as was noted long before Descartes came along) Mind.

94 We've already noted subtle evidence for this when considering the implications of Bell's inequality (see endnote 84), as well as Schrödinger's cat (see endnote 131)—not to mention when closely attending to the colors of a sunset or the hoots of a late-night owl.

95 There are potentially unlimited theodicies. That fact alone ought to be enough to signal ANYONE that they're on the wrong track. As in an endless game of whack-a-mole, when each new defense is mounted, each, in turn, is shot down. And there can never be any final defense for what is bla-

tantly defenseless. It's not any particular set of tit-for-tat that needs to be dropped. It's the very exercise itself, mired in assumptions of substantiality on both sides, that must be *seen* as utterly futile.

96 There's a Taoist story of a wise Chinese farmer whose horse ran away. His neighbor came by to console him on his bad luck. "Who can say what's good or bad?" said the wise farmer. The next day the farmer's horse returned, bringing several other horses with her. The neighbor paid a visit to congratulate the farmer on his good fortune. "Who can say what's good or bad?" said the farmer. The next day the farmer's son broke his leg trying to break one of the new horses. The neighbor came by, once again, to commiserate with the wise farmer. "Who can say what's good or bad?" said the farmer. The next day the army came by conscripting young men for the war, but they passed over the farmer's son because of his broken leg. Again, the neighbor came to applaud the farmer's good fortune. "Who can say what's good or bad?" said the wise farmer.

There's no end to this story.

97 Decisions like this ought to be left to the many. There is wisdom in it, provided that the many are reasonably well-informed. Leaving such matters to the few, though they may be rich and powerful, wrings wisdom out of the decision making, and disaster normally follows. Wisdom never functions among partisans, but only arises out of the Whole.

98 If you're interested in physics: this is where our problems with Schrödinger's cat come from. Likewise, for all the other quantum conundrums attendant to this elementary conceptual mistake. We'll settle this in appendix B, part 2, and in endnote 131.

99 From *You Have to Say Something* by Dainin Katagiri, p. 153.

100 Even though I use the term "they," I'm not speaking of a single group of people. Groups of people were scattered in many directions from the central area where the Great

Flood took place. These groups carried elemental dialects of their common language—proto-Indo-European—in all directions, radiating outward from their home territory, which is now under the Black Sea.

101 For a more complete and scientifically based description of what appears to have occurred during the Great Flood event, see appendix C: The Flood and Other True Fictions.

102 As we noted in chapter 19, though we seem to parse out distinctions—indeed, this is what conscious experience is—upon close investigation, we can't find actual boundaries. Thus, it's Awareness of Wholeness that correctly guides us in living a morally upright life, rather than fighting over which moral principles are best among the distinctions that endlessly appear to rise and fall away.

103 Our habit of seeing good as set against evil has probably been with us since time immemorial. But it was Zoroaster who pulled these two into clear contention when he personified them into Ahura Mazda, the embodiment of Good, and Angra Mainyu, or Ahriman, the embodiment of Evil. Zoroaster envisioned human history as an absolute choice—you are either with Good and against Evil, or you are the reverse. Or, in other words, you're either with us (you're Good) or against us (you're of the Devil). This extreme dualistic view eventually infected the thinking of Jewish leaders while they were held captive in Babylon for 40 years, roughly four centuries before the time of Jesus. From there, such thinking went on to infect the traditions of Christianity and Islam, and eventually most of the Western world.

Today this contagion of thought spreads worldwide. But it's not the only way to view things. Eastern traditions, for example, generally place importance on balancing opposites into harmonized wholes—as depicted, say, in the yin-yang symbol associated with Taoism.

These are two very different ways of understanding the World and our place in it. One is Wholesome, for It views everything in terms of Wholeness, or Totality, while the other views everything in terms of vying entities. And while the first remains calm and all-embracing, the latter is ever driven by fear and remains at war—even as it dreams of total victory. Yet there can be no such outcome for those who ignore or dismiss Wholeness.

104 Mystics typically invoke mystery and thereby stop short of *waking up*. But there is actually no Mystery (with a capital M), as we will see.

105 This awareness—that we don't find or experience a persisting self—is one of two profound insights attributed to the historical Buddha. (The other insight, which is closely related, has to do with willed action and its link to dissatisfaction, confusion, and suffering. I discuss this in some detail in my book *Buddhism Plain and Simple*. In particular, please read the footnote on pages 156–57 of that volume.)

106 If you're interested in physics: Rebirth consciousness is akin to explanations of how changes occur in quantum fields. They "propagate in a 'laborious' manner, with a change in intensity at one point causing a change at nearby points," explains physicist Rodney Brooks in his book *Fields of Color* (Rodney A. Brooks, 2010). Like so, in a manner similar to dropping a stone in water, as the disturbance moves outward from the point of impact, the disturbance at any given point affects adjacent points. Such transmitting disturbances, of course, are not merely disturbances in space, but disturbances in time as well. In fact, it is such disturbances that make this moment appear very much like, but not identical to, what appears to have occurred "adjacent to" or immediately "before."

107 The term actually associated with reincarnation—and that was on everyone's lips at the time of Gotama—would be literally translated as "redeath," not "rebirth." Contrary to the way people in the West regard the notion of reincarnation today, the people of Gotama's time and culture

regarded the prospect of relentlessly dying, over and over again, with dread. They sought a way to be released from redeath. Gotama pointed out to them that what they feared was actually a mistaken view. Because of the subtlety of his message, however—i.e., that we don't actually find a self, a persisting being that is subject to death, let alone to redeath—very few people, even to this day, have gotten it right. Of course, this needs to be *seen* directly, not figured out, or believed.

108 As, down through the ages, many other awakened teachers did not claim to belong to any particular sect.

109 Shunryu Suzuki, in his classic Zen text *Zen Mind, Beginner's Mind* wrote, "I discovered that it is necessary, absolutely necessary, to believe in nothing. That is, we have to believe in something which has no form and no color—something which exists before all forms and colors appear." And the ancient Græco-Roman philosopher Sextus Empiricus, of course, wrote, "For it is sufficient, I think, to live by experience, and without subscribing to beliefs . . ."

110 This understanding has to do with Totality. The equivalency I'm pointing to here becomes obvious when Reality is *seen* as Mind.

111 History is replete with examples of discarded hypotheses—e.g., phlogiston, caloric, and the æther, once thought to be necessary to fill all of space. Using the method of science, over time such poor ideas get weeded out as better, sturdier, and more workable hypotheses come along. This is not a weakness of science, but arguably its greatest strength. Science is self-correcting. Few institutions devised by humans have this remarkable attribute.

112 While science **must** form hypotheses, religion has no such need.

113 The Sanskrit term Nagarjuna used was ***shunyata***, which is typically (and inadequately) translated into English as "emptiness," or sometimes as "voidness." These terms, however, are problematic in that they give the mistaken

impression that *shunyata* refers to Nothingness, à la Leibniz. What *shunyata* actually refers to is not Nothing, but to a lack of substantiality. Thus, I've rendered the term here as *nonsubstantiality*.

114 This is not to say that there are no "small m" mysteries—everyday mysteries that appear in our conceptual world, such as where I left my keys, or what became of my umbrella.

115 Science **can** venture somewhat into nonconceptual aspects of Reality via pure mathematics, but such excursions are akin to driving blind. In these regions, consciousness provides no conceptual constructs for science to follow. For more, see appendix B, part 2 (a discussion on physics and consciousness). For more from a purely mathematical perspective, turn to the discussion on quaternions in appendix D, item 6.

116 There is a great deal more that can be said here. Nature can productively tinker with, say, genetics because it operates as the Whole. There's Wisdom in it. We, on the other hand, work from our very limited and self-centered perspective, and inevitably come up with dreadful oversights that have the potential to bring about disaster—including our own extinction. Blaise Pascal put it well when he wrote, "All human evil comes from this, man's being unable to sit still in a room."

117 Science has recently gained this powerful tool. *Symmetry*, as a scientific principle, is not merely mirror symmetry, or rotational symmetry, as you might be picturing. It refers to any transformation within a system that leaves the overall system unchanged. What scientists have come to realize over the last century is that, if this principle does not appear within their mathematical models, then there's a good chance that their hypotheses are wrong. Symmetry, as is now realized, is foundational to all physical laws. Symmetry also appears to be a deep, underlying principle of Nature. We'll return to symmetry later in this book.

118 The famous photo, titled "Earthrise," was taken on the Apollo 8 mission in December 1968.

119 The Hubble Deep Field is a breathtaking photo taken by the Hubble Space Telescope. It confirmed that there are millions of galaxies in every direction, in each minuscule patch of sky.

120 This has led some to proclaim that there **is** no Ultimate Truth. If this includes you, please turn to appendix A, where we will together examine the nature of Truth in some detail.

121 The word "religion" is rooted in the Latin *religare*, which means "to bind again" or "to bind strongly"—by implication, with what is ultimately Genuine, Real, and True.

Originally, religion was probably nothing more than performing rituals in order to divine the future, win battles, get better yields on crops, produce healthy children, ensure a prosperous life, and so on. But since the Axial Age, during the first millennium BCE, many religions began to range far beyond such provincial concerns, looking deeply into what it means to be human.

122 This was the response I received from a monk when I asked about these images upon entering a Zen temple in Japan. I've never been able to confirm these names, but the impression was powerful and had a profound effect on my understanding of Zen practice from that moment on.

123 This is to mistake a vision of the Absolute for Enlightenment. *Seeing* Oneness is no more than realizing the nonsubstantiality of all things. It is realizing that mountains are not mountains and rivers are not rivers. Significant as this might seem, it is still not Enlightenment.

124 If you, the Reader, have more questions for me, please go to mtsrmts.com.

125 Even though Edmund Gettier shot down the currently accepted base definition of Knowledge as a "justified true belief" in 1963, philosophers are loath to part with it. Virtually every epistemologist knows there's something wrong

with this definition, but apparently, they haven't a clue as to what it is. Thus, they keep fiddling with it—adding things to it or redefining their assumptions in various ways. Yet nothing ever works. And nothing **will** work, given their reluctance to venture outside of the conceptual.

The problem, once again, is belief. Calling it "justified" or "true" doesn't help. Nor does asserting additional considerations, such as taking into account the "virtuous motives" of the believer, or applying additional notions like "undefeated justification," or any other sort of tinkering philosophers have drummed up over the years. There can **be** no "justified true belief," no matter how much we tweak the idea. Mistaking belief for actual Knowledge will never, under any circumstance, magically yield Knowledge.

I say a bit more about the "Gettier problem," as this is called, in my book *Why the World Doesn't Seem to Make Sense*, p. 46.

126 There are indeed other truth theories, though ***pragmatism*** and certain varieties of ***coherence theory*** are the most notable. All truth theories, however, presume substantiality—and, thus, all are useless and mistaken from the get-go.

127 See endnote 96.

128 There are people who think that we humans are incapable of solving this problem because our brains lack the capacity to process the information that would be required. Yet the resolution (but not the solution, since such a flat-earth problem has no "solution") does not demand much of our intellects. Others think the problem is unsolvable because we can't get outside of consciousness to study it. It's true that we can't get outside of consciousness—but we don't **need** to. Rather than fruitlessly trying to analyze what can't be analyzed, we only need to *see*.

129 These are references to well-known conjectures, assumptions, FEQs, and discussions that assume without question that Mind is a product of matter.

130 The way in which AI people think about brains and minds—comparing them to computers—borders on the silly. Just because solving a simple problem, such as determining what is wrong with a photograph of an inverted tiger walking across a living room ceiling, might seem to require "having lots of contextual knowledge, vastly more than can be supplied with the algorithms that advanced computers depend on to identify a face or detect credit-card fraud," it doesn't mean that beefing up the computing power and data loads of computers is finally going to do the trick. **What goes on in a human mind is nothing like what goes on in a computer.** And it doesn't even have to be a human mind. Even a bird, with its little bird brain, can immediately distinguish the difference between a dog and a cat. Yet this is a feat that "super-smart" Watson, and even so-called "deep learning" techniques, continue to struggle with (see "Machines Who Learn," by Yoshua Bengio, *Scientific American*, June 2016, pp. 46–51). The quote used above is from "A Test for Consciousness—How will we know when we've built a sentient computer? By making it solve a simple puzzle," by Christof Koch and Giulio Tononi, *Scientific American*, June 2011, p. 46.

131 If you're interested in physics: In 1935, Erwin Schrödinger, annoyed by the absurd implications of quantum mechanics, devised his famous thought experiment involving a cat in order to bring the impossibly strange peculiarities of the quantum world into the everyday world we all take for granted. He showed how seemingly legitimate yet ambiguous states belonging to particles can, without resorting to any sort of magic, be extended into the everyday world of people and cats. Thus, he showed that, just as you can have, say, an electron in some arbitrary superimposed state of, say, emitting/not emitting a photon, you could rig an experiment in such a way that would put a cat into a superimposed state of being alive/dead at once.

Much hullabaloo has been raised over this issue (Stephen Hawking once complained that whenever he hears about Schrödinger's cat, he reaches for his gun), but none of the hypotheses put forth to settle the matter have ever successfully done so.

One of the more recent attempts, **Quantum Bayesianism** (or **QBism**, for short), basically asserts that the superposition in question doesn't apply to an "objective world" "out there"—i.e., to photons and cats, but merely to the "observer's subjective mental state" (for more, see "Quantum Weirdness? It's All in Your Mind," in *Scientific American*, June 2013, pp. 47–51). The problem with this particular attempt is that the "observer's mental state" is neither separate, nor separable, from "the observed." In other words, Schrödinger's thought experiment points out that there is neither a separate "objective world" "out there" nor a "subjective mental state" somewhere else.

What physicists have yet to hit upon is that it's our staunch and never-questioned belief in substantiality that is the problem. In other words, we've never actually had an objective, separately formed cat in the first place. Nor do we **ever** have any such formed entities, including subatomic particles, or "yourself." As Schrödinger's thought experiment directly hints at, these are all manifestations of Mind.

Thus, when it comes to cats, or stars, or people, or atoms, or quarks, we can't really say that there are any of these things. Can't really say that there are not, either. We just simply never get to actual particularness. In other words, we never get to Something.

132 "It has in it the heart of quantum mechanics. In reality, it contains the only mystery." Richard Feynman, from his *Lectures on Physics* and *The Character of Physical Law*.

133 Einstein made this remark in 1951, though shortly before his death in 1955 he embraced quantum field theory.

134 Since confirmation of the Higgs field, however, many par-

ticle physicists are willing to admit that there aren't any particles—things, if you will—only fields. Yet particle physicists still hold onto the idea of particles, if only because it makes everything easier to work with, mathematically speaking.

135 For example, a photon—i.e., a quantum of light—is not a formed entity, a thing, an object, a particle, but a disturbance traversing the electromagnetic field. An electron is not a formed thing, but a disturbance traversing the electron field. A graviton is not a formed entity, but a disturbance traversing the gravity field. And so on.

136 The problem is that a "point particle" seems to momentarily signal at either one hole or the other—either that, or at neither hole (as if "it" were absorbed at the screen, or somewhere else). In other words, the unacknowledged problem for the field physicist is field collapse—which, as we're about to see, field physicists regard as a mystery or gap in their knowledge.

137 I highly recommend *Fields of Color* as an excellent introduction to quantum field theory (QFT) for the lay reader. Here, in plain English, with the use of color, and without challenging the reader with mathematics, Dr. Brooks shows how the weird and paradoxical oddities of quantum mechanics (QM) are resolved in QFT. Not only that, he also shows how QFT, unlike QM, is quite compatible with relativity. In fact, relativity theory is easily derived from QFT. Dr. Brooks leaves you with the feeling that QFT is so sensible and successful that you may wonder why anyone would resist embracing it.

Then, for a good counter read to *Fields of Color*, though a more technical one, I suggest "What Is Real?" by Meinard Kuhlmann, in *Scientific American*, August 2013, pp. 40–47. Here Dr. Kuhlmann lays out not only the trouble with particles, as we've already seen in the main text of this book, but the trouble with fields as well. He then offers a couple of possible extensions of QFT that may resolve these troubles, bringing you to the frontiers of this research.

138 Particle physicist Sean Carroll, in a 2012 TAM talk. Other particle physicists have echoed Carroll's observation.

139 The remaining three gaps Brooks identifies have to do with (1) what physicists call *renormalization*, (2) why masses and field strengths are what they are, and (3) dark matter and dark energy. These three additional gaps are tangential to our immediate discussion, so I will not take them up here. As mysteries, however, they, too, will fall away with a proper understanding of immediate, direct experience.

140 Quotes are from *Fields of Color* by Rodney A. Brooks, chapter 8, under the heading "Field collapse."

141 Field collapse is simply a "point particle" signaling "its" location or some other property. In other words, a distinction is being made—which, of course, is precisely what consciousness is. (See endnote 136.)

142 For more, see "What constitutes a measurement?" in item 8 of appendix D.

143 If you're still wondering what sentience or conscious awareness is, I suggest that you not rush by this point.

So, let's slow down.

I just wrote, "if **you**'re wondering." Our language forces it. But here's a hint: **this** is where the problem is. It's not with conscious awareness. Conscious awareness is unreservedly direct.

The problem has to do with "you." (Or with "cat," if you want to apply this to Schrödinger's famous thought experiment.) It's in our deep-seated but unjustified assumptions of particularity where we get everything wrong.

144 The veiled and erroneous assumption here is that the term "someone" (or, "something," or "cat") has an actual referent in Reality.

145 This is not to say that the **idea** of a quantum field is inconceivable, but only to say that "what goes on in a quantum

field" is not available to conscious experience. As we're about to investigate, it involves what we literally can't imagine.

146 If you're **really** interested in the physics: In a classical field, every point in spacetime can be assigned a specific value— as in "a charged particle in this location at this moment feels x amount of electrical force." Not so in a quantum field. In a quantum field, each point in spacetime is not assigned a value (a number designating something able to be pictured), but an "operator," such as, say, "$\sqrt{}$" ("square root"). If at this point you're saying "huh?" to yourself, don't worry. You're not alone. As I mentioned before, you can't picture what's going on in a quantum field. Physicists can't, either. They're at a total loss in trying to understand what particles or fields even are.

The mathematical abstraction above, however, doesn't end with ungraspable operators. You actually **can** arrive at a graspable value (an *eigenvalue*) for any given point in a quantum field, but you must first apply the operator to yet another mathematical entity called a *state vector*.

Never mind what that is. Never mind what any of this is, really. The point that concerns us here is that the state vector, unlike an operator, is **not** specific to any particular "location," but refers to the system as a Whole.

In other words, once the operator is applied to a state vector, we are no longer dealing with itty-bitty entities appearing in time and space, but with Totality. **This** is why we can't picture It. **It** is not an object. It's purely Mind.

147 A reminder: Symmetry, as used both in this book and in physics, refers to any dynamic transformation within a system that leaves the overall system unchanged. For example, if you rotate a square 90°, the square will appear just as it did before the transformation.

The Pythagorean theorem also shows symmetry. In all right triangles, no matter how much the lengths of the shorter sides change, the sum of their squares always equals the square of the hypotenuse (the longest side opposite the 90° angle).

Over the last century, physicists have come to realize that if such symmetries do not appear within the equations describing their hypotheses, they know they're on the wrong track. Symmetry has come to be regarded as one of the foundational characteristics of the laws of physics. It is woven into the fabric of Reality.

148 Nagarjuna, the great sage from second-century India, wrote: "Those who do not understand the distinction between these two truths cannot understand what the Buddha taught." From Nagarjuna's *Verses on the Middle Way (Mulamadhyamakakarika)*, chapter 24, verse 9, my translation.

149 This event, known as the Younger Dryas, left its mark through drastic changes—not just in the northern hemisphere, but around the globe, even as far away as New Zealand—as the Earth plunged back into near-glacial conditions for the next 1,200 years.

150 This was glacial Lake Agassiz. Though you can read about it elsewhere, I give a brief history of it in my book *Why the World Doesn't Seem to Make Sense*.

151 As with many of the scenarios constructed from archeological and geological evidence, some of the details are controversial. Some sources say the inflow was equivalent to a thousand Niagaras. Others put it at 100. Most say the basin filled over a matter of days or weeks; others say it took a month or more. One source says a year, which seems unlikely. But, while the details will continue to be debated, what is certain is that a great and sudden flood did create the Black Sea at a time when many of the flooded lands were inhabited.

152 To the ancients, the sky was a tangible, physical place that you could walk around in (if you could get to it)—a bit like a forest canopy. It was nearby, but in a direction not easily traversed by mortals.

153 We've been trained to think that The Flood resulted from a rain that lasted forty days and forty nights. But stories of such rains were likely later embellishments added to tales as they were handed down, generation after generation. The people who witnessed The Flood described the waters as coming from below.

Thus, in many myths and stories from that era, two kinds of waters are often mentioned. There are abyssal waters beneath, as well as waters above the firmament, as described in Genesis, and by Thales, the forerunner of all Greek philosophers. Thales, no doubt picking up on general impressions people had for generations after the time of The Flood, claimed water to be the source of all things. He also had the idea that the world floated about on a vast extent of water, as an earthy clod.

There are other quaint stories of violent, flooding waters that date from this same time and place—just as, from another time and place, there are stories of a giant frog who spewed water upon the land. Each meshes well with what we know of the experiences of people who witnessed such events.

154 This was the Meluhhan civilization, as the Sumerians (who traded with them) called them. They flourished around 2500 BCE, possibly earlier. Today we refer to them as the Harappan civilization, after a modern-day town in the region. They were a very advanced culture, with paved streets, plumbing, and public baths. They seem to have been a very peaceful people as well, since few weapons appear among their artifacts, or in depictions of daily life in their artwork. They also seem to have been familiar with early versions of many of the gods later associated with Hinduism. Otherwise, though, we know little about them, since we can't read their writings. It may very well have been these people who first made use of decimal numbers.

155 Proto-Indo-European eventually spread into twelve language families: Indic (including Sanskrit, Pali, and its

other descendants), Iranian, Anatolian (an extinct group, including Hittite and other languages), Armenian, Hellenic (Greek), Albanian (though it is possibly descended from Illyrian), Italic (including Latin, Italian, Spanish, Portuguese, French, and Romanian), Celtic, Tocharian (an extinct language group from central Asia), Germanic (including English, German, Dutch, Gothic, Norwegian, Swedish, Danish, and Icelandic), Baltic, and Slavic (including Russian, Czech, Bulgarian, Serbian, and Croatian).

156 To put it differently, calculus is effective in describing motion precisely because persisting entities (i.e., illusory thought constructs) can't move. In other words, calculus deals only with snapshots, however quickly we would imagine a shutter opening and closing.

157 Quaternions are complex numbers that take the form of w + xi + yj + zk, where w, x, y, z are "real" numbers and i, j, k are "imaginary" components.

While "real" numbers—consisting of integers (whole numbers), rational numbers (those that can be expressed as ratios between wholes, such as ½, or –¾, or as repeating decimals), and irrational numbers, such as π, or $\sqrt{2}$ (expressed as non-repeating decimals)—can have conceptual referents, "imaginary" numbers, such as i—or the square root of –1 ($\sqrt{-1}$)—have no conceivable referents whatsoever.

158 This argument says that if certain constants of nature were ever-so-slightly different from what they are now—say, by one point in the $10^{-125\text{th}}$ decimal place—then atoms, stars, and galaxies would never have formed. The universe would have either collapsed long ago, or it would have been blown to smithereens. And, of course, we wouldn't be here to contemplate this amazing coincidence, or design, or feat of fine-tuned cosmic engineering.

159 The *many-worlds hypothesis* proposes that the universe splits into two with each and every particle interaction within each and every quantum system—which is both

absurdly extravagant and supported by a total of zero evidence and zero experience.

For more on the many-worlds hypothesis see "The Quantum Multiverse" in *Scientific American*, June 2017, pp. 28–35, where the problem of other universes is linked to the many-worlds conjecture. For more on the topic of other universes in general, see, "Does the Multiverse Really Exist?" in *Scientific American*, August 2011, pp. 38–43.

ABOUT THE AUTHOR

 Steve Hagen has been an instructor in religion at St. Olaf College and a science researcher for the University of Minnesota and the State of Alaska. In 1979 he was ordained a Zen priest, and in 1989 he received formal endorsement to teach. He has, however, no formal ties to any Zen or Buddhist organizational hierarchy. In 1997, he founded Dharma Field Meditation and Learning Center in Minneapolis, where he served as head teacher until 2021, and where he continues to speak and lead classes. He is the author of the bestselling *Buddhism Plain and Simple* and several other popular books on religion, science, and philosophy.

Ending the Pursuit of Happiness
A Zen Guide
Barry Magid

"This book provides a radical and vitally important challenge to the prevailing cultural ethos."
—Jeremy D. Safran, PhD, editor of *Psychoanalysis and Buddhism*

Emptiness
A Practical Guide for Meditators
Guy Armstrong

"For anyone seeking to understand emptiness, this is a clear and fine guidebook, with precise and practical ways to explore and deepen your practice."
—Jack Kornfield, author of *A Path with Heart*

Being-Time

A Practitioner's Guide to Dogen's Shobogenzo Uji

Shinshu Roberts

"This book is a great achievement. Articulate, nuanced, and wonderful."

—Jan Chozen Bays, author of *Mindfulness on the Go*

**Science and Philosophy
in the Indian Buddhist Classics**

Volume 1: The Physical World

Conceived and introduced by H. H. the Dalai Lama

Edited by Thupten Jinpa

"A rare gift of wisdom from the ancient world to the modern reader."

—Daniel Goleman, author of *Emotional Intelligence*

The Zen of R2-D2

Ancient Wisdom from a Galaxy Far, Far Away

Matthew Bortolin

"Finding spirituality in Star Wars can bring that galaxy far, far away a lot closer to home."

—*Newsweek*

About Wisdom Publications

Wisdom Publications is the leading publisher of classic and contemporary Buddhist books and practical works on mindfulness. To learn more about us or to explore our other books, please visit our website at wisdomexperience.org or contact us at the address below.

Wisdom Publications
199 Elm Street
Somerville, MA 02144 USA

We are a 501(c)(3) organization, and donations in support of our mission are tax deductible.

Wisdom Publications is affiliated with the Foundation for the Preservation of the Mahayana Tradition (FPMT).